D0955646

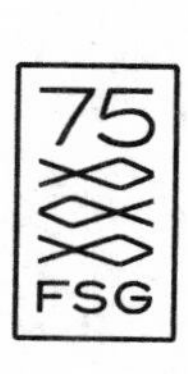
75
FSG

# THE QUICK FIX

# THE QUICK FIX

## Why Fad Psychology Can't Cure Our Social Ills

**Jesse Singal**

FARRAR, STRAUS AND GIROUX / *New York*

Farrar, Straus and Giroux
120 Broadway, New York 10271


Printed in the United States of America
First edition, 2021

Portions of chapters 1 and 6 were adapted from articles originally published on the website of *New York* magazine.

Library of Congress Cataloging-in-Publication Data
Names: Singal, Jesse, 1983– author.
Title: The quick fix : why fad psychology can't cure our social ills / Jesse Singal.
Description: First edition. | New York : Farrar, Straus and Giroux, [2021] | Includes bibliographical references and index. | Summary: "How popular psychology fails to solve problems facing society and draws attention and resources away from more effective structural fixes"— Provided by publisher.
Identifiers: LCCN 2020049200 | ISBN 9780374239800 (hardcover)
Subjects: LCSH: Psychology—Popular works. | Human behavior. | Social problems. | Psychology.
Classification: LCC BF145 .S535 2021 | DDC 150—dc23
LC record available at https://lccn.loc.gov/2020049200

International Paperback Edition ISBN: 978-0-374-60408-0



1 3 5 7 9 10 8 6 4 2

*To my parents, Sydney Altman and Bruce Singal,*
*who gave me all the love and support I needed and then some,*
*who taught me there are no quick fixes, and who let me*
*skip school once in a while just for the hell of it*

# CONTENTS

# THE QUICK FIX

# INTRODUCTION

If you're the sort of person who buys and reads books about human behavior, then it is likely you have recently encountered an exciting, counterintuitive new psychological idea that seems as if it could help solve a pressing societal problem like educational inequality, race relations, or misogyny. Maybe you came across it in a TED Talk. Or, if not there, in an op-ed or blog post or book.

It is, after all, a golden age for popular behavioral science. As the University of Virginia law professor Gregory Mitchell, a keen critic and observer of the field, wrote in 2017, "With press releases from journals and universities feeding a content-hungry media, publishing houses looking for the next Gladwellian bestseller, governments embracing behavioral science, and courts increasingly open to evidence from social scientists, psychologists have more opportunities than ever to educate the public on what they have learned about why people behave as they do."[1]

I should know. Starting in March 2014, I was the editor of Science of Us, *New York* magazine's newly launched online social science section. It was my job, and the job of the very talented people I worked with, to find new, interesting behavioral science research to write about every day of the week and to do so in a rigorous, sometimes skeptical manner.

Thanks to a fairly stats-heavy master's program I had completed, I knew some of the differences between good and bad research, and some of the ways quantitative claims can mislead. What I didn't anticipate was the fire hose of overhyped findings that would fill my email in-box daily, the countless press releases from research institutions touting all-caps AMAZING results that would surely blow my mind, and the minds of our readers, bringing us impressive web traffic in the process. I treaded water as best I could, trying to resist the lure of bad science by writing and editing stories we could be proud of. But I don't think I quite grasped the full scale of the problem.

That changed in September 2015, six months or so into my job. That was when I met with Jeffrey Mosenkis, who does communications for the research organization Innovations for Poverty Action and who also happens to have a PhD in comparative human development (a blend of anthropology, social psychology, and other fields). Perhaps because he had noticed that I appreciated the debunking of overhyped research findings, he had decided to pass along a tip: I should look at the flaws in the implicit association test.

Maybe you've heard of this test. Commonly referred to as the IAT, it is seen by important people with impressive credentials as *the* most promising technological tool for attenuating the impact of racism. The idea is that by gauging your reaction time to various combinations of words and images—say, a black face next to the word "happy," or a white one next to the word "danger"—the test can reveal your unconscious, or implicit, bias against various groups. The test, introduced in 1998, has been a blockbuster success. Anyone can take it on Harvard University's website, and over the years its architects and evangelists, some of the biggest names in social psychology, have adapted it for all sorts of diversity-training uses. It would be hard to overstate its impact on schools, police departments, corporations, and many other institutions.

In his email, Jeff said there was a story waiting to be told by a journalist; the test was weak and unreliable, statistically speaking. A group of researchers had published work showing rather convincingly that the test barely measures *anything* of real-world import. Which, if true, would naturally raise some questions about the possibility that Harvard was apparently "diagnosing" millions (literally millions) of people as unconsciously biased on the basis of a flimsy methodology, and about all the money being spent on IAT-based trainings.

I was intrigued by Jeff's email and soon got in touch with one of the skeptical researchers, Hart Blanton, who was then at the University of Connecticut. As I started looking into his claims, I realized that I had simply assumed the test did what its most enthusiastic proponents said it did, despite the rather audacious nature of their claim: that a ten-minute computer task with no connection to the real world could predict subtle forms of real-world discrimination. I had credulously accepted these claims because I had figured that if almost the entire discipline of social psychology had embraced this innovation as a cutting-edge tool in the fight against racism, and a multitude of organizations outside academia had followed suit, all these people must have known what they were doing. This brought a pang of shame. "I believe this thing because a lot of people say it is true" is not a great stance for a science writer and editor.

After a lot of reading and interviewing, I concluded Blanton was correct. As I wrote in a subsequent article, the statistical evidence for the sorts of claims the IAT's creators were making was sorely lacking.[2] There was a gap between what many people believed to be true about the test and what the evidence revealed. Part of what was going on was that the IAT told a good story that lined up with certain liberal anxieties about race relations, tinged with an optimistic note—*Sure, everyone's racist deep down, but this test can help us discover and undo those biases, one person at a time!*—and people wanted to believe that story. If millions of people believed in the IAT largely on the basis of good storytelling and the impressive credentials of its creators, what did that say about the manner in which we talked about these issues? Could it be that urgent tasks like police reform were being approached in the wrong way? And what about the broader current state of behavioral science? What else was I missing?

WHAT I SOON REALIZED is that our society's fascination with psychology has a dark side: many half-baked ideas—ideas that may not be 100 percent bunk but which are severely overhyped—are being enthusiastically spread, despite a lack of hard evidence in their favor. The IAT is one example, but there are numerous others. The popularity of these ideas, as well as the breathless manner in which they are marketed by TED Talks and university press offices and journalists and podcasts, is not harmless. It misallocates resources to overclaiming researchers when others are experiencing a funding crunch, and it degrades the institution of psychology by blurring the line between behavioral science and behavioral pseudoscience.

Perhaps most important, these ideas are frequently being adopted by schools, corporations, and nonprofits eager to embrace the Next Big Thing to come out of the labs and lecture halls of Harvard or the University of Pennsylvania. As the decision-makers who work in these institutions have grown more fluent in science and cognizant of the need to look to behavioral scientists for guidance (a good thing), they've also become more susceptible to half-baked behavioral science (a bad thing). And this explosion of interest in psychological science has occurred at a time when, if anything, people should be *more* cautious about embracing new and exciting psychological claims. As we'll see, many findings in psychology—including those featured in introductory textbooks—are failing to replicate, meaning that when researchers attempt to re-create them with new experiments, they are coming up short. This so-called replication crisis has cast a giant shadow over the entire field of psychology, and the best available evidence suggests that a sizable chunk of published psychological findings may be false (though the size of that chunk is a source of heated debate). Many psychologists themselves may be unwittingly helping to promote half-baked science that *seems* to be built upon a solid foundation of published research. In all likelihood, we'll look back wincingly at some of the popular theories being taught and developed today and, all too often, transformed into sleek interventions that promise to alleviate our ills.*

THIS BOOK IS AN ATTEMPT to explain the allure of fad psychology, why that allure is so strong, and how both individuals and institutions can do a better job of resisting it. It is important to improve our understanding of how behavioral scientific information circulates in the public realm, because only sound knowledge will earn us the improvements we wish for. Just as we can't enact successful environmental and energy policies while denying global warming, and

* To be sure, many other fields of research, including cancer biology, are experiencing their own replication crises. While we'll learn more about psychology's replication crisis in chapter 7, readers interested in how these issues have affected other fields should consider reading Stuart Ritchie's great 2020 book, *Science Fictions: How Fraud, Bias, Negligence, and Hype Undermine the Search for Truth*. And Daniel Engber's 2016 piece in *Slate* "Cancer Research Is Broken" offers an excellent primer on the travails of that particular field.

we can't improve global public health without taking a stand against anti-vaccination myths, we will never solve the pressing social issues of the day—racism and inequality and the education gaps and so many others—while relying on claims about human behavior and how to change it that are half-true at best.

The spread of half-baked behavioral science can't be explained apart from the present state of American political and intellectual life. The country has suffered from decades of rising inequality paired with interminable political dysfunction, and as institution after institution has seen its legitimacy crumble, there's been an ever-intensifying focus on the individual. We're living in what the Princeton historian Daniel Rodgers calls an age of fracture, the title of his invaluable 2011 book. "Conceptions of human nature that in the post–World War II era had been thick with context, social circumstance, institutions, and history gave way to conceptions of human nature that stressed choice, agency, performance, and desire," he explains. "Strong metaphors of society were supplanted by weaker ones. Imagined collectivities shrank; notions of structure and power thinned out."[3] In this dispensation, we are taken to be discrete individuals floating around in markets, increasingly responsible for our own well-being and increasingly cut off from the big groups and institutions and shared ideas that gave American life so much of its feeling and texture and meaning in the past. (Of course, many Americans, by dint of their race or gender or religion or sexual orientation, would not want to return to that past.)

Americans are also living with the consequences of what the political scientist Jacob Hacker has termed "the great risk shift": as the nation's social safety net has frayed and ever more risk has been off-loaded from companies and the government onto overburdened Americans, economic insecurity has crept higher and higher up the income and wealth ladders. This only exacerbates the sense that everyone must fiercely defend their gains and stand vigilant against the possibility of sliding down into a less advantaged position. The combination of the age of fracture with the great risk shift likely affects what sort of behavioral science wins out in the marketplace of ideas: it likely shifts the focus toward improving and optimizing and repairing individuals rather than understanding how they are influenced by big, roiling forces largely beyond their control.

Within psychology, particularly social psychology, these tendencies have given rise to what I call Primeworld,[4] a worldview fixated on the idea that people's behavior is largely driven—and can be altered—by subtle forces. Central to Primeworld is, well, "primes," the unconscious influences that, according to some psychologists, affect our behavior in surprisingly powerful ways: holding a warm drink makes people act more warmly toward others, claims one finding, and being exposed to stimuli connected to the elderly makes people walk slower, claims another. The proponents of Primeworld suggest we can work toward "fixing" individuals by helping them to understand the influence of primes and biases. Their accounts have three main characteristics: big, imposing social structures and systems are invisible, unimportant, or improved fairly easily; primes and biases have an outsize influence on societal outcomes; and these primes and biases can be fixed, to tremendously salubrious effect, thanks to the interventions offered by wise behavioral scientists.

This worldview treats complicated problems as though they can be significantly ameliorated or solved with quick fixes: with cute, cost-effective interventions by psychologists. Over and over, throughout this book, we will see situations in which otherwise brilliant researchers examining some of the most complicated problems known to humankind have adopted the tenets of Primeworld. Whereas a social scientist taking a broader view might look at a problem like "the education gap" between white and black students and explain that it has many complicated causes, ranging from segregated schools to early-life experiences to the impact of tutoring, Primeworld adherents will take a different tack. They will stress that a great deal of progress can be made by optimizing the individual participants in the system, whether students (perhaps via increasing their "grit," meaning their ability to stick with difficult problems tenaciously) or teachers (perhaps via IAT-based trainings).

The point is not that Primeworlders deny that there's a bigger world out there, beyond primes and biases; if you asked them, they would quickly acknowledge that yes, there is. The problem is that their work speaks for itself, advancing a set of very specific, very zoomed-in priorities. As a result of their evident excitement over exploring biases and primes and the potential to optimize individuals, almost everything else fades into the background, indistinct and all too easily ignored. But it is increasingly clear that Primeworld simply hasn't delivered. Over and over, it has been shown that the interventions its members favor simply don't warrant the hype they generate, and there's strong reason to believe that they fail *because* they neglect to attend to deeper, more structural factors that are not easily remedied via psychological interventions.

As this book will show, this does not mean we should give up on applied behavioral science entirely. Psychology has produced some ideas—particularly certain so-called nudges and interventions targeting institutions rather than individuals—that have held up well under scrutiny, and is likely to produce more as methodological reforms take hold (which, as we will see, has already begun happening).

Still, there are good reasons to be wary of the styles of psychological research that have seized public attention in recent years. And yet we fall for quick-fix, half-baked behavioral science over and over again.

# 1

# THE SELLING OF SELF-ESTEEM

What I remember is the image of a balloon. I'm in kindergarten or first grade, seated on the floor with other tiny people, and a teacher is explaining to us that we all have this balloon in us called self-esteem—a point she is illustrating with an actual balloon. Sometimes, people are nice to us or we do well in school, and that feeling inflates the balloon a little; here I don't remember but can only assume that the teacher blew into the balloon. Other times, bad things happen—we're excluded or get mad or get yelled at—and the balloon deflates (the teacher lets a little air out). The more air that is in the balloon, the better off we'll be, we are told. We're likely to make smarter decisions, act more kindly toward others, and simply be better people overall. These balloons are very important.

This would have been in 1990 or so, and all around the country other children in other classrooms were hearing the same thing. The general tenor of the era's obsession with self-esteem is captured rather memorably in a 1991 children's book called *The Lovables in the Kingdom of Self-Esteem*. Written by Diane Loomans and illustrated by Kim Howard, *The Lovables* imparts a simple, nurturing message: you, the child reading this book or having this book read to you, are *very* special.

The inside copy reads as follows:

> I AM LOVABLE!
> I AM LOVABLE!
> I AM LOVABLE!
> By using these magical words, the gates to the Kingdom of Self-Esteem swing open for readers of all ages. Inside the Kingdom live twenty-four animals—the Lovables—each one with a special gift to contribute. Mona Monkey is lovable. Owen Owl is capable. Buddy Beaver takes care of the world around him. Greta Goat trusts herself.

*The Lovables* is no outlier. If you grew up or raised a child during the 1980s or 1990s, you likely remember such messages. A certain ethos took hold during this time: self-esteem was just about the most important thing society could give a young person—a crucial trait that could mean the difference between success and catastrophe—and therefore it was vital for schools to impart it.

Though it wasn't just for schoolkids. At the peak of this fad, just about *everyone*, from CEOs to welfare recipients, was told—often by psychologists with serious credentials—that improving their self-esteem could "unlock the gates" of success. This was both a personal argument and a political one: the movement, which had its epicenter in California, claimed that increasing people's self-esteem could reduce crime, teen pregnancy, and a host of other social ills, even pollution. And once it caught on, it was taken quite seriously by policy makers, to the point where numerous states would allocate taxpayer money specifically toward self-esteem programs.

The vogue for self-esteem had many origins, but its institutionalization in schools was mostly the work of a single, very eccentric California politician. John Vasconcellos, who died in 2014, was a Democratic state legislator representing Silicon Valley for thirty-eight years—thirty in the state assembly and eight in the state senate. In his obituary, the *San Jose Mercury-News* described him as a "famously rumpled bear-of-a-man" who was "colorful, witty, brilliant, angry, intellectual and elegantly foul of mouth."[1] Most of all, though, he was a nonconformist—during one three-year stretch, he decided to just let his hair grow and grow and grow—and his nonconformity frequently took on a decidedly California hue. Vasconcellos was an idealist who was convinced that humans had untold, untapped greatness, but it was an idealism driven in part by a bevy of personal demons and a long-running battle to control his anger problems. He was quite public about his varied attempts at self-improvement, which ranged from obscure forms of therapy to the teachings of the New Age Esalen Institute in Big Sur.

Vasconcellos, known as Vasco around Sacramento, was a searcher in every sense of the word. "By the mid-1980s," writes the British journalist Will Storr in his book *Selfie*, "Vasco's intellectual explorations had taken him to far shores, and he'd become notorious, around the Capitol, for some of his crazier ideas. He wondered, for example, if a special form of 'gentle birthing' might lead to less violent humans. He was apparently at least curious about the idea that children having sex with their parents was not, in fact, harmful to the child but natural and healthy; he'd read reports of women expressing pride at having lost their virginities to their fathers and invited a proponent of paedophilia to join his 'network on sexuality.'"[2]

Along the way, likely through his Esalen contacts, Vasconcellos discovered the writings of the psychotherapist Nathaniel Branden, a disciple (and lover) of Ayn Rand who, like Rand herself, was fixated on the importance of self-esteem. Branden's 1969 book, *The Psychology of Self-Esteem*, sold more than a million copies.[3] Branden could not have been clearer or more forceful in his assessment of self-esteem's importance. "There is no value-judgment more important to man—no factor more decisive in his psychological development and motivation—than the estimate he passes on himself," he wrote.[4] He later asserted that he could not "think of a single psychological problem—from anxiety and depression, to fear of intimacy or of success, to alcohol or drug abuse, to spouse battering or child molestation, to suicide and crimes of violence—that is not traceable to the problem of a poor self-concept."[5]

Branden's book was mostly polemical and anecdotal, but by the time Vasconcellos came across it, there was also a body of psychological work that appeared to back it up. Generally speaking, these studies would measure a group of people's self-esteem, measure some sort of behavioral or cognitive outcome, and gauge how closely the two were correlated. Self-esteem was measured with a variety of self-report instruments, the most popular one being the Rosenberg Self-Esteem Scale (first published in 1965), which asks respondents to rate their agreement with items like "At times I think I am no good at all" and "I am able to do things as well as most other people" on a four-point scale.[6]

Some studies found that people with high versus low self-esteem reacted in better, more adaptive ways to various challenges. The influential social psychologist Roy Baumeister, who has studied the subject, told me that self-esteem seemed to correlate with "how favorably [subjects] react when we tell them they failed on the first trial of a [multi-trial] task." The higher a study participant's self-esteem, the more likely they were to stick with the task in question. Other published findings appeared to demonstrate a link between self-esteem and important real-world outcomes: the higher someone's self-esteem, the better their academic and work performance, for example.

Advocates and researchers, then, were able to point to a body of work and treat it as evidence that self-esteem was an important causal factor in people's success. As we will soon see, that literature is in fact significantly more muddled than it appears at first glance. Nevertheless, Branden's book and the findings appearing to support it were career-defining discoveries for Vasconcellos. The logic was simple: if low self-esteem is tied to so many maladaptive responses, to so many forms of underachievement and bad behavior, then surely raising kids' (and others') self-esteem could bring with it untold benefits. Soon, Vasconcellos was lobbying Sacramento, and his longtime political nemesis, the Republican governor, George "The Duke" Deukmejian, to launch a statewide commission to study the public policy implications of self-esteem. After being stymied by a number of legislative false starts and near misses—and by a massive heart attack, after which Vasconcellos wrote to his constituents "asking them to imagine themselves with tiny brushes swimming through his arteries, scrubbing at the cholesterol"[7]—he got his commission: in 1986, Governor Deukmejian signed legislation creating the California Task Force to Promote Self-Esteem and Personal and Social Responsibility. The task force's goals were to explore how to apply self-esteem to a range of social problems, and the state budgeted it $245,000 per year. Naturally, it was headed by Vasconcellos.

Much of the nation laughed in response. Garry Trudeau pilloried the concept regularly in his *Doonesbury* comic strip, where his ditzy, New Agey character Boopsie was named to a fictionalized version of the task force. (In one later strip, the panel is declared a success by its members, who announce, in creepy unison, "WE FEEL GOOD ABOUT FEELING GOOD ABOUT YOURSELF!" "Excuse me," Boopsie says. "Hunk-Ra, the ancient warrior for whom I channel, would like to dissent."[8]) National ridicule or no, the task force's members set to work. And its members weren't just from the healing-crystals set; according to one *Washington Post* article, the group consisted of "a mix of fundamentalist Christians, gay activists, law enforcement officers, educators, counselors," and, yes, "New Age believers." They all were confident about their new project. In fact, as

*The Post* paraphrased it, they predicted "that many other states will follow suit once they see that the idea can save the state money spent on truant officers and prison cells."[9]

The task force didn't get off to a particularly quick or efficient start. It took its members more than a year to come up with a working definition of self-esteem in the first place: "appreciating my own worth and importance, and having the character to be accountable for myself and to act responsibly toward others" eventually won out. But soon the group gained a bit of momentum, and the underlying idea seemed to become increasingly accepted by the public: What if all sorts of bad social outcomes really *were* caused directly by low self-esteem? That would suggest a very easy vector for fixing them—one that could appeal not only to liberals with countercultural sympathies but to budget-conscious conservatives, too. This bipartisanship was essential to the idea's success and widespread adoption. In one Associated Press interview, for example, a Republican member of the task force framed self-esteem initiatives as an alternative to "pouring money down the so-called rat holes" of traditional approaches to fighting social problems. "We keep spending money on crime and violence," he said. "We keep spending money on drug abuse."[10] If government could find cheap ways to instill higher self-esteem, the thinking went, it could obviate the need for much of that spending.

The idea of self-esteem as a panacea caught on not only because of these policy promises and the scientific evidence that seemed to support them but also because it tapped into what was, by then, a long-standing American belief that simply adopting the right mindset can have positive (and, in some tellings, miraculous) results. "Happiness depends more on the inward disposition of mind rather than on outward circumstances," wrote Ben Franklin in *Poor Richard's Almanack*, recounting what he described as the wisdom of ancient philosophers.[11]

Perhaps the most interesting strain of this belief arose in the late nineteenth century, in the form of the so-called New Thought movement. While New Thought was a complicated amalgamation of many ideas, including what was then called Oriental philosophy, it owed its greatest debts to Phineas Parkhurst Quimby, a clockmaker and mesmerist who mentored and "healed" the Christian Science founder, Mary Baker Eddy, and to Ralph Waldo Emerson and the other American transcendentalists. But it was sui generis: as the Yale historian Alfred Whitney Griswold, a critic of the movement, described it in 1934, it was "a system of high-powered mental telepathy which held that matter could be spiritualized and brought under the complete domination of thought, and, conversely, that all thoughts become matter." Naturally, America being America, this telepathy was mostly put toward the goal of personal prosperity. "The great majority [of the movement's adherents] were in it for what they could get out of it, and that was money," wrote Griswold. "They wanted to succeed, to grow rich, to rise in the world, rather than to commune with the All-Mind." This idea, that people's own thoughts contained tremendous untapped power, imbued New Thought with "its promise to make [its adherents] persons in an impersonal world."[12]

New Thought was metaphysically zanier than the self-esteem movement, but some of its offerings bear a striking similarity to that later fad. "The mood of success should be held constantly in mind in all the work of this book, and in all the affairs of life," wrote the New Thought leader Frank C. Haddock in *The King's Achievements; or, Power for Success Through Culture of Vibrant Magnetism*, which was published in 1903 and which sat in his "Power-Book Library" alongside titles like *Power of Will*, *The Personal Atmosphere*, and *The Culture of Courage*. "This may be done by affirming, until it is a permanent belief and expectation of the soul, 'I am resolved on success. I shall certainly achieve success.'"[13] ("I AM LOVABLE! I AM LOVABLE! I AM LOVABLE!")

New Thought's heyday didn't last long—Griswold was already writing about it in the past tense in 1934—but some of its core ideas continued to percolate. The movement's "intellectual and spiritual heir," as the historian Daniel Horowitz puts it, was the minister Norman Vincent Peale. While Peale toned down some of New Thought's wilder flights, the general message remained intact. His famous 1952 bestseller, *The Power of Positive Thinking*, promises in its *first sentence* that "you do not need to be defeated by anything, that you can have peace of mind, improved health, and a never ceasing flow of energy." All by adopting the right outlook.

Peale's book offers anecdote after anecdote supporting his central claim, and skeptics are easily swept away by the power of his ideas. "I suggested these principles some months ago to an old friend of mine, a man who perpetually expects the worst," writes Peale at the end of a chapter titled "Expect the Best and Get It." "He expressed vigorous disbelief in the principles outlined in this chapter and offered to make a test to prove that I am wrong in my conclusions." But after he followed Peale's suggestions, which centered on "one of the most powerful laws in the world . . . change your mental habits to belief instead of disbelief," he realized he was mistaken. "I am now convinced," he later told Peale (or according to Peale he did, at least), "although I wouldn't have believed it possible, but it is evidently a fact, that if you expect the best, you are given some strange kind of power to create conditions that produce the desired results."[14]

Peale's descendants are everywhere. In the first decade of the twenty-first century, Oprah Winfrey came under fire by some for promoting *The Secret*, the popular book by Rhonda Byrne that was premised on the "law of attraction," or the idea that "what you think about, you bring about." It can't be overemphasized how literally Byrne, who has made many millions of dollars off *The Secret*, intends this point to be taken by her audience: one of the testimonials on her website is written by a woman who visualized her dream car and, not long after, got that very car (with the help of a zero-down-payment loan).[15] The website also helpfully explains that *The Secret* is a scientifically backed idea. "Under laboratory conditions," after all, "cutting edge science has confirmed that every thought is made up of energy and has its own unique frequency."[16]

So the idea that positive thinking can improve one's lot stretches over American history like a comfortable blanket. But as Vasconcellos's Esalen-tinged spiritual journey suggests, the self-esteem movement was also very much of its time. During this period, bestselling books like *I'm OK—You're OK* (1967) focused on the principle that people have deep psychic wounds that need to be addressed before they can fully actualize themselves. By the time the self-esteem hype started to build, it is likely that many readers of this literature had deeply internalized the belief that their self-conception problems were holding them back but could be fixed. The self-esteem craze, then, can be seen as the confluence of two powerful currents in American cultural life: the long-standing belief in the power of positive thinking and the more recent belief that people needed to address their deep psychic wounds. In short, *Yes, you are broken, but if you start to feel better and more positively about yourself, you can be fixed.*

A lot of this may come across as mystical or mawkish, but it's worth noting that by the time Peale burst onto the scene, many researchers and social theorists had investigated and theorized about self-esteem in a more serious-minded way. "A man with a broadly extended empirical Ego, with powers that have uniformly brought him success, with place and wealth and friends and fame, is not likely to be visited by the morbid diffidences and doubts about himself which he had when he was a boy," wrote William James in *The Principles of Psychology: Volume 1* in 1890. "'Is not this great Babylon, which I have planted?' Whereas he who has made one blunder after another, and still lies in middle life among the failures at the foot of the hill, is liable to grow all sicklied o'er with self-distrust, and to shrink from trials with which his powers can really cope."[17]

Vasconcellos's task force, then, was able to build upon a fairly substantial cultural edifice. But it also did the PR work of generating colorful anecdotes of its own to buttress its cause: it held a series of events around California in which police officers, social workers, ex-cons, and others testified to the importance of self-esteem. Yes, there were embarrassing hiccups—like the time a photographer caught the commission's members holding hands in a circle after their lunch break—but overall the group was surprisingly successful at carving out a place in the national conversation about drugs, crime, and other social ills.

There were also complicated institutional reasons the self-esteem craze caught on and had such a long run. We know the details thanks to the work of the British writer Will Storr, who spent a year digging through archives in California and interviewing many of the key players. In *Selfie: How We Became So Self-Obsessed and What It's Doing to Us*, he explains that after the self-esteem commission was formed, Vasconcellos was able to persuade the University of California system to convene a panel of its own researchers to study self-esteem and to publish the results through the University of California Press.[18] That effort was coordinated by Neil Smelser, a professor emeritus of sociology at UC Berkeley. The release of these findings, everyone understood, would be "of critical importance" to Vasco and his movement: either the experts would support their stratospheric claims about the importance of self-esteem, or they wouldn't. Vasconcellos and his team would find out at "7:30 p.m. on 8 September 1988 at the El Rancho Inn in Millbrae," just south of San Francisco.[19]

At this point the story branches off in two different directions. In the official version, the news they got at the El Rancho Inn confirmed to Vasconcellos and his colleagues that they were on the right track; according to their research, self-esteem mattered a great deal. That's what Vasconcellos and company told the world in a report they made public a few months later. An AP write-up explained to readers that the UC researchers had "[lent] legitimacy to the commission's founding premise: Poor self-esteem is closely linked with alcoholism, drug abuse, crime and violence, child abuse, teenage pregnancy, prostitution, chronic welfare dependency and failure of children to learn."[20] A pile of other news coverage imparted the same message: hey, make fun of this stuff if you want, but now we know that it's legitimate science; that's what the scientists themselves say.

Then there's what really happened at the meeting. Through interviews and audiotapes and other records he unearthed, Storr found out that Smelser had *not*, in fact, told the task force that the science backed up their claims. Rather, he had presented a much more conflicted view of the research. In a tape discovered by Storr, Smelser can be heard saying of one slice of the self-esteem research that "these correlational findings are really pretty positive, pretty compelling"—a quotation that made it into the subsequent media coverage—but quickly pivoting to also say that "in other areas the correlations don't seem to be so great and we're not quite sure why." With regard to alcoholism, he asks, "Do these people go to drinking because of an earlier history of self-doubt, self-degradation, worthlessness and so on? Or is it the other way around?"[21] That is, maybe they feel worthless *because* they drink so much. At the end of his recorded remarks, Smelser makes his opinions clear: it would be "the sin of overselling" to tell California's legislature that the UC research had come back in a manner that clearly and straightforwardly buttressed the case for self-esteem. "And nobody can want to do that," he told the group. "You don't want to do that. Certainly, we don't want to do that."

At the time, these more critical words didn't make it out of that room at the El Rancho Inn. Instead, the task force, in summing up the meeting, only quoted the "really pretty positive, pretty compelling" bit, distorting the true sense of Smelser's remarks and falsely portraying him as standing firmly behind the self-esteemers' claims. According to David Shannahoff-Khalsa, a member of the task force who was so put off by the process he withheld his signature from the final report, this was intentional obfuscation. Vasconcellos and the other true believers knew that accurately presenting what the UC researchers had found could curtail the state's appetite for funding self-esteem initiatives. "Oh, it was absolutely dishonest," Shannahoff-Khalsa told Storr.[22] It was "a fucking lie."[23] (In Storr's book, this quotation isn't attributed to anyone by name, but Storr confirmed to me it was Shannahoff-Khalsa who uttered it.)

Why did Smelser allow himself to be used in this manner? As it turned out, "he'd been forced into playing a highly delicate political game," as Storr puts it.[24] Smelser, who died in October 2017, explained to Storr that he needed to balance his desire for UC to publish sound science with his need to mollify Vasconcellos, an extremely important and easily angered cog in the complicated machine that was Sacramento.

As head of the state government's Ways and Means Committee, Vasconcellos controlled the UC budget purse strings. If Smelser piped up too much—if he said, *Hey, that wasn't what I said at that meeting!*—it could bring his institution a lot of grief. So he trod a middle ground: in private he told the task force the truth, but in public he stayed quiet after his message was muddled and willfully misinterpreted by the self-esteemers.

One couldn't come up with a more revealing example of how the scientific ideal of disinterested objectivity can diverge from reality. In some purer realm than the one we actually inhabit, Smelser's path would have been clear: Come out swinging against Vasconcellos's misrepresentation. Explain, loudly and repeatedly, that the science did *not* show what the task force claimed it did. But Smelser had one set of incentives—scientific integrity—pointing him one way, and another—the politics of public-university funding—pointing him elsewhere. "I think my work was successful, in that regard," he told Storr, "because the university was never criticized."[25] A year after that meeting at the El Rancho Inn, the UC group published its book, *The Social Importance of Self-Esteem*, with Vasconcellos and Smelser listed as two of the three editors. Its conclusions were underwhelming, according to Baumeister. "Boy, they did not find very much," he said. It also didn't sell that many copies.[26]

BUT IN THE LATE 1980s, none of this intriguing backstory was public, and none of it mattered. What mattered was that the fictionalized version of the 1988 meeting won out, and it helped spur the publication of the task force's own report in 1990, *Toward a State of Esteem*,[27] which took a more optimistic view of the research and, naturally, left a much larger cultural imprint. "*Toward a State of Esteem* went on to be a victory that lay far outside the reasonable hopes of anyone who'd witnessed its humiliating origins," writes Storr. "The Governor of Arkansas, Bill Clinton, who'd privately mocked Vasco and his project, now publicly endorsed it, as did serious figures including Barbara Bush and Colin Powell. Although, naturally, it could never have claimed to have convinced every member of the American press, the majority bent knees towards it."[28]

Even before the official release of this report, the credulous media coverage spread. Glowing anecdotes and confident quotations received little pushback. "I've read hundreds of files on criminals," California's special assistant attorney general Brian Taugher told a United Press International journalist. "Some people end up in a life of serious crime simply because they dislike themselves."[29] Perhaps further confusing matters, contemporaneous reports frequently conflated self-esteem interventions with other things. For example, a 1989 *Baltimore Sun* article described a San Jose "curriculum based on principles of self-esteem": "The framework, which applies to teachers as well as students, is based on developing a sense of security, identity, belonging, purpose and personal competence. A student who is doing poorly in geography, for instance, has the opportunity to take a learning partner in class, to work with an older student or a teacher, or to take an after-school study hall."[30] But giving a student extra academic help isn't boosting their self-esteem, exactly, at least not directly; it's giving them extra academic help.

Vasconcellos and his colleagues, too, had a tendency to describe all sorts of social programs as falling under the umbrella of "self-esteem," even rather traditional examples of social and academic support services that didn't target self-esteem directly. There was some mission creep here, and it confounded matters, because even if these programs did improve outcomes, that wouldn't necessarily be due to their self-esteem components. This is another feature of half-baked ideas we'll see pop up again and again: they tend to spread outward, like a gas, filling up any empty space in the conversation. As an idea without a firm basis and clear conceptual boundaries gains more and more purchase in the public imagination, there's less and less rhetorical precision in how it's used. And because there are then more opportunities for it to be invoked, it will subsequently spread even further.

The result of all this excitement over self-esteem was an increasingly massive cottage industry. It's difficult to estimate that industry's true size—the broader self-help industry, of which self-esteem is a part, brought in about $10 billion in 2016[31]—but a *New York Times* article from 1990 captures its scope. "Hundreds of school districts have added self-esteem motivational materials to their curriculums," wrote the reporter Lena Williams. "American employers have turned increasingly to consultants who say they can raise employees' morale and work performance through self-esteem techniques. New companies have formed, devoted to teaching on self-esteem themes, and hundreds of books on self-esteem and self-enhancement have been published."[32]

The most successful self-esteem entrepreneurs raked in millions. Jack Canfield, the founder of the Santa Barbara–based Self-Esteem Seminars, offered self-esteem seminars involving videos, audiotapes, and kinesthetics (Canfield is also the author of the mega-bestselling *Chicken Soup for the Soul*, which launched a veritable empire of sequels and offshoots). In some states, welfare recipients were given workbooks designed to help them boost their self-esteem, and of course companies had to produce those materials, too. The cottage industrialization of self-esteem might have further disincentivized many people—not just the peddlers of self-esteem materials, but the school-district and corporate decision-makers who had already shelled out money for them—from viewing the concept with too skeptical an eye.

The craze hit American schools harder than just about any institution. And once it did, it produced an endless assortment of colorful classroom interventions. One common exercise for grade-schoolers involved a Koosh ball. A kid tosses the ball to another kid and compliments him: *I like your shirt.* Then that kid tosses the ball to someone else and compliments her: *You're good at soccer.* The good feelings travel with the Koosh ball across the room, back and forth and back and forth. This is somewhat similar to the "Magic Circle" exercise described in a 1990 *Globe and Mail* account of a Toronto classroom:

> It's 9:30 a.m., Magic Circle time in Room Six at Winchester Public School.
>
> A dozen third-graders and their teacher, Oksana Hohol, sit cross-legged on an old rug. Ms. Hohol welcomes each child. Today's topic, she says, is "something nice I have done for a friend."
>
> They think for a few minutes. Lydia puts her hands together, signalling that she wishes to speak. "Some kids were picking on one of my friends, so I gave her a big hug," she says.
>
> Other children describe similar good deeds. They praise each other. Oksana, as the children call their teacher, thanks each student by name and later asks what they like about "Circle."
>
> "I feel good when I share my feelings," one child says.[33]

Other schools stopped using red pens, the theory being that seeing a lot of red on a spelling test could harm a child's self-esteem. Some installed mirrors with text like "You are now looking at one of the most special people in the whole wide world!" emblazoned on them.[34]

Surely some of these activities, particularly those geared at five-year-olds, who wouldn't exactly have been learning calculus otherwise, were harmless. But according to the social critic Steve Salerno, certain concepts that came out of the self-esteem craze transformed education for the worse. In his polemical but well-researched book *Sham: How the Self-Help Movement Made America Helpless*, he shows that it wasn't just Koosh balls and cheesy mirror exercises; in many schools, prevailing assumptions about academic rigor and feedback changed too. The thinking went, "Don't make kids feel bad about everything, because if they feel bad they'll perform poorly," as Salerno told me. Self-esteem also established itself, in an arguably counterproductive way, in the long-running national conversation about societal inequality. "There was this sense of the inner city falling behind—specifically black kids in the inner city are not performing as well as other kids," said Salerno. "And there was this assumption that it was because they lacked self-esteem." If you can boost their self-esteem, you can close the achievement gap. The nice thing about this theory, Salerno noted, is it doesn't require much of a fundamental reworking of the educational system; it's something of an easy way out. In many cases, advocates focused on self-esteem "rather than hiring better teachers, spending more money on actual schools and instruction. It became a surrogate for the stuff that might actually have done some good." In future fads, the pattern would repeat: the reforms that ask the least of us are often the ones most apt to go viral.

AS IT TURNED OUT—and as the mostly forgotten *Social Importance of Self-Esteem* suggested—there was very little validity to the causal claims everyone was making about self-esteem in the 1980s and 1990s. We know this with a fair degree of certainty because around the turn of the century, long after self-esteem programs had blossomed all over North America, the psychological establishment decided to take a more critical, in-depth look at the extant research on this subject. Roy Baumeister and three other researchers were invited by the American Psychological Society to conduct a comprehensive review of the literature with the goal of finding out whether self-esteem really "works" as advertised.

Baumeister was initially, like so many others, a believer in the straightforward importance of improving self-esteem and had published some studies of his own in the field. But he began observing the movement with increasingly arched eyebrows. "Starting in '84 and '85," he told me, "I started to watch for contrary evidence and began to notice that things aren't as rosy as I had assumed." The claims Vasconcellos and others were making were just too big and extravagant; nothing they said was as carefully hedged as it should have been, and they rarely, if ever, acknowledged certain gaps or inconsistencies in the available data that were apparent to Baumeister. For one thing, the self-esteem evangelists were claiming that raising self-esteem would lower crime rates, but Baumeister had found in some of his research that criminals actually had *higher* self-esteem than law-abiders.[35] For another, the project was motivated, at least in part, by the frequently echoed claim that America was suffering from an "epidemic" of low self-esteem, and yet when Baumeister and his colleagues had investigated that question in 1989, they'd found the opposite: most Americans had high self-esteem, at least according to the tools researchers used to measure it.[36]

Baumeister teamed up with fellow psychologists Jennifer Campbell, Joachim Krueger, and Kathleen Vohs, and the fruit of their efforts was a very important 2003 article in *Psychological Science in the Public Interest* titled "Does High Self-Esteem Cause Better Performance, Interpersonal Success, Happiness, or Healthier Lifestyles?" It was paired with a general-interest article in *Scientific American* two years later,[37] and it convincingly debunked most of the claims about self-esteem that had fueled the craze.

Advocates for the self-esteem movement had, for decades, been treating the scientific literature on the subject as though it told a simple story about the obvious importance of self-esteem in causing various outcomes. But Baumeister and his colleagues treated the literature far more skeptically, almost as though it were a storage locker filled to the brim with sundry old trinkets and knickknacks and furniture, the sort of place where a trained eye is required to sift the useless junk from the valuable antiques.

Such an eye is needed, a lot of the time. We'd like to think it's otherwise, that scientific progress is straightforwardly linear, that every study adds something to our knowledge. And yet many published studies just *aren't very good*; they end up pointing in the wrong direction because they aren't carefully constructed, their authors don't benefit from methodological advances that come to seem obvious just a few years later, or for various other reasons.

Reading Baumeister and his colleagues' study offers a strangely potent thrill if you are fascinated by social science and the many ways it can go awry. At root, it's simply a review of the available literature on self-esteem. But the execution of that review is so clearheaded, so careful, so *scientific*, that a document ostensibly about a rather dry subject ends up reading more like a detective story.

Perhaps their most important finding was that relatively minor-seeming differences in how self-esteem was studied appeared to yield massive differences in what was subsequently discovered, particularly when it comes to the distinction between objective and subjective measures of performance. In attempting to correlate self-esteem with school performance, for example, do you simply ask students how they do in school, or do you evaluate their actual grades? While the answer might seem obvious—of course it's better to look at someone's actual grades than what they *say* their grades are—that isn't always practical for a given researcher. What if you can't find a big group of students who will hand over evidence of their grades? It's much easier to simply ask them to rate their own performance—it cuts out an entire annoying step in the research process—which is the sort of shortcut some researchers had taken over the years. To paraphrase Donald Rumsfeld, sometimes you study psychology with the data you have, not the data you want.

People with high self-esteem, perhaps unsurprisingly, have a tendency to rate themselves highly in various domains of life, often in a reality-defying manner that would put them right at home in Garrison Keillor's Lake Wobegon, "where all the women are strong, all the men are good-looking, and all the children are above average." This effect, it turned out, seriously distorted some of the findings linking self-esteem to positive outcomes. In 1993, for example, one researcher, summing up past studies, found that—as Baumeister and his colleagues put it—"people's physical attractiveness accounted for more than 70% of the variance in their self-esteem."[38] If true, this would be an astounding result, and a depressing one for those of us who turn few heads upon entering a party, because it would imply that most of what makes us feel good or bad about ourselves has to do with our looks. Luckily, studies measuring this correlation in a more objective manner, by having third parties rate people's attractiveness based on photographs rather than accepting self-reported attractiveness at face value, reached a different result: attractiveness accounted for, at best, just 2 percent of the variation in people's self-esteem, leaving a lot of room for the influence of other factors.*

---

* "Variation explained by" references how tightly two variables appear to be linked. In the models being discussed here, factors *unrelated* to attractiveness account for 98 percent of the differences in people's self-esteem levels, dwarfing the importance of attractiveness in explaining these differences. There is genuinely no way to explain this concept in an entirely intuitive, nontechnical manner, but the key thing to keep in mind is that the higher the percentage, the more one variable appears to explain another. The variation in people's height in centimeters accounts for 100 percent of the differences in their heights in inches, for example, because those two variables are just different ways of measuring the same thing. That is, if you measure someone's height in inches, it tells you exactly what their height in centimeters is, whereas if you measure their attractiveness, it gives you almost no useful information about what their self-esteem is likely to be. Generally, statistical models employing these concepts include more than two variables—later in the book, for example, we'll

Plainly put, this is the difference between self-esteem mattering and not really mattering much at all. And "over and over during our survey of the literature," wrote Baumeister and his colleagues, "we found that researchers obtained more impressive evidence of the benefits of self-esteem when they relied on self-reported outcomes than when they relied on objective outcomes."

To be fair, in some cases, Baumeister's team was left with small or moderate correlations linking self-esteem even to *objectively* measured positive outcomes. One example is the "positive but weak" correlation that remained between self-esteem and school performance. But the existence of such a correlation (weak or otherwise) isn't, on its own, proof that self-esteem is important in the way Vasco and so many others claimed it was. Their key claim was not just that self-esteem correlates with certain positive outcomes but that it *causes* them. This gets to a classic social science issue: the difference between correlation and causation, and the fact that the former does not imply the latter. As Baumeister and his colleagues put it, "A correlation between X and Y could mean that X causes Y, that Y causes X, or that some other variable causes both."

Taking the example of Y causing X rather than vice versa, imagine that I tell you I've made an alarming discovery: Playing in the NBA causes people's heights to sprout up suddenly, as evidenced by the fact that the average height of an NBA player is about six feet seven inches, or about nine inches taller than the average American male.[39] Such a claim would likely cause you to revise your estimate of my intelligence downward, because of course it isn't being in the NBA that causes people to be more likely to be tall, but the reverse: being tall makes it more likely you'll enter the NBA. That's an extreme and obvious example, but in many instances even otherwise

---

encounter research in which intelligence, grit, physical fitness, and other variables all account for different amounts of the variation observed in the likelihood of military cadets successfully traversing a grueling training regimen.

careful researchers can be fooled into thinking X causes Y when in reality Y causes X.

Baumeister and his colleagues' meticulous tour of the literature strongly suggested that good grades might (rather weakly) cause higher self-esteem, not the reverse, perhaps because getting A after A after A causes a student to begin to have more and more faith in her abilities. In one study, for example, a pair of Norwegian researchers "found evidence that doing well in school one year led to higher self-esteem the next year, whereas high self-esteem did not lead to performing well in school."[40] Even here, though, the evidence was mixed; other carefully conducted studies hadn't found much of a correlation at all.

In other instances, "some other variable" causes both X and Y to move in unison. When researchers fail to identify this variable, instead assuming X causes Y, this is known as "omitted variable bias," or "third variable bias." Let's say that I discover that in the Massachusetts town where I grew up, there is a statistically significant correlation, in the winter, between the weather being cloudy on a given day and the public schools being closed that day. From this I conclude that cloudiness sometimes causes schools to close. This is an incorrect causal claim, because I'm omitting the third variable that actually causes the closures: snow. Clouds cause snow (so they're correlated), and snow causes school closures (ditto), but clouds don't directly cause school closures, so those two variables end up being correlated despite the absence of a *causal* link.

Baumeister and his colleagues found evidence that omitted variable bias accounted for certain self-esteem findings. As it turned out, the few researchers who had carefully controlled for these so-called confounds had been finding this all along. In one study from 1977, for example, researchers had concluded that "shared prior causes, including family background, ability, and early school performance, affect self-esteem and later educational attainment and were responsible for the correlation between the two."[41] Of course, if you don't measure these other variables and carefully account for them in your statistical analysis, you might "discover" a straightforward-seeming causal relationship between self-esteem and school performance, in much the same way if you failed to account for snow, you might decide that clouds cause school closings.

Baumeister and his colleagues also found that along the way some self-esteem researchers had used "path analysis," "a statistical technique for testing theories about complex chains of causes."[42] Path analysis is designed to shed more light on the likelihood of causal influence than the discovery of mere correlations between variables can. In theory, it allows research to more confidently advance claims of the sort "A and B are correlated with each other, and they each *cause* C to go up or down, which in turn *causes* D to go up or down."

When researchers brought this tool to bear on the self-esteem question, they tended to come up empty-handed. In one study, "there was no direct causal path from self-esteem to achievement." In another, "the direct link[s] from high school self-esteem to later educational attainment . . . indicate that the relationship [was] extremely weak, if it exists at all."[43]

On the whole, "Does High Self-Esteem Cause Better Performance, Interpersonal Success, Happiness, or Healthier Lifestyles?" is a wonderfully detailed forensic analysis of why so many people had been fooled into believing certain claims about self-esteem, and it is genuinely useful for anyone hoping to understand not only the self-esteem controversy but the issue of half-baked scientific findings more broadly. Baumeister and his colleagues' literature review shows that there were many warning signs, often in the form of published papers from decades prior, that certain causal and correlational claims for self-esteem were likely being overstated.

In one sense, though, all of this was moot. Despite the explosion of self-esteem programs, Baumeister and his colleagues "found relatively little evidence on how self-esteem programs or other interventions affect self-esteem" in the first place. Decades after the start of the craze, there was simply a dearth of solid research on this fundamental question, and because many of the programs in existence "target[ed] not only self-esteem but also study skills, citizenship, conflict reduction, and other variables," it was difficult to interpret the results in a way that isolated the role of self-esteem.[44] So even if there had been a clear causal relationship in which self-esteem caused (say) school performance, there weren't any proven interventions to boost the former.

Baumeister and his colleagues were forgiving of the psychologists and others who had contributed to the self-esteem craze. "Was it reasonable to start boosting self-esteem before all the data were in?" they wrote. "Perhaps. We recognize that many practitioners and applied psychologists must deal with problems before all the relevant research can be conducted."[45] This is, in fact, a common occurrence in psychological science: You have a handful of papers pointing to a correlation that *could* have important real-world ramifications, assuming certain other things are also true. But it takes a while to determine whether those certain other things are true, and in the meantime other people—people who might not be as committed to scientific rigor as the best social scientists are, or who are trying to solve urgent real-world problems and don't have the luxury of waiting for more peer-reviewed evidence to come in—might decide to run with the idea before a genuinely trustworthy verdict arrives.

That seems to be what happened here. Despite the absence of a truly robust base of causal evidence linking self-esteem to positive outcomes, it was an irresistible story. From the point of view of excitable politicians like Vasconcellos, either the idea was self-evidently true and there was little need for hard evidence anyway, or there was *enough* evidence to go ahead and run with it. For influential would-be brake appliers like Smelser, there were incentives to not be the sole naysayer in the room. And while there were other skeptics—including the conservative social commentators Charles Krauthammer and Dr. Laura Schlessinger,[46] who saw the self-esteem movement as yet another manifestation of the saccharine, mushy self-help drivel that was, in their view, undermining American toughness—they were easily drowned out by all the enthusiasm.

That's why a simple, highly viral message—raising self-esteem can greatly improve people's lives and productivity—was able to catch on, offering a straightforward solution to a constellation of problems that are not, in fact, straightforward to solve.

SO IS THAT THE END of the story? Should we simply brush aside the idea that self-esteem is important as totally misbegotten, as complete pseudoscience? No: that, itself, would be a bit of an oversimplification.

It's true that there is little evidence for a clear-cut, causal relationship between self-esteem as captured by psychologists' tools and the various life outcomes Vasconcellos was concerned with, and, for that matter, a lack of interventions proven to boost self-esteem over the long run. But there does appear to be *some* truth to the idea that when people get too down on themselves, it can cause harm and hinder their performance and happiness, and that it's possible to prevent this from happening with certain types of efforts.

A good portion of the evidence supporting these beliefs comes from the related ideas of mindset theory and cognitive behavioral therapy, or CBT. Mindset theory, pioneered by Carol Dweck of Stanford University, posits that people's attitudes and attributions regarding certain types of events are greatly affected by whether they have a *growth mindset* or a *fixed mindset*. In school settings, where these ideas have been studied most thoroughly, those who have a growth mindset believe intelligence and performance can be improved over time; when they fail at a task or get a poor grade, they attribute that setback to a lack of practice or experience and view the negative outcome as useful feedback and an opportunity to improve and do better next time. Those with a fixed mindset, on the other hand, believe that intelligence and skill flow mostly from innate ability and can't really be improved through practice. They interpret failure differently: *I got a D because I'm bad at math; that's all there is to it.*

Dweck has long claimed that certain interventions can effectively shift people, particularly students, from fixed to growth mindsets and that their performance improves as a result. There's a strong case to be made that she and other proponents of the idea have overclaimed in certain ways. "When you enter a mindset, you enter a new world," she wrote in her book *Mindset: The New Psychology of Success*, suggesting that learning about mindsets can cause the reader to "suddenly understand the greats—in the sciences and arts, in sports, and in business—and the would-have-beens. You'll understand your mate, your boss, your friends, your kids. You'll see how to unleash your potential—and your children's."[47]

The evidence does not support these and other lofty claims that both Dweck and others have made about mindset interventions over the years. But in 2019 the biggest, most robust study of these interventions was published in *Nature*, and it did show some promise, at least in specific groups, from a cost-benefit perspective. A team of twenty-five researchers, including Dweck, randomly assigned members of a sample of 12,490 American public school ninth graders to a two-part intervention geared at instilling a growth mindset (each half was just twenty-five minutes long, making for fifty minutes of total time). The intervention "communicates a memorable metaphor: that the brain is like a muscle that grows stronger and smarter when it undergoes rigorous learning experiences," explain the authors. "Adolescents hear the metaphor in the context of the neuroscience of learning, they reflect on ways to strengthen their brains through schoolwork, and they internalize the message by teaching it to a future first-year ninth grade student who is struggling at the start of the year."[48]

Now, an obvious difference between this approach and self-esteem curricula is that many of the latter simply told kids they *were* good or smart, whereas mindset interventions tell kids they can *become* those things. But still, this mindset intervention targets a similar trait and does appear to have some effect. Overall, the intervention "reduced by 3 percentage points the rate at which adolescents in the U.S. were off-track for graduation at the end of the year," though it didn't seem to have much of an effect on stronger students not at risk of falling off-track.[49] If this research holds, it could be argued that mindset interventions do offer a minor but legitimate boost to a subset of otherwise academically vulnerable students—a boost that is at least somewhat related to self-esteem.

Cognitive behavioral therapy (CBT), which dates back to the work of the psychiatrist Aaron T. Beck in the 1960s, has a stronger evidence base and also seems to tap into some of the same forces Vasconcellos was interested in. Practitioners hold that suffering results not just from what happens to a person but from how a person interprets the meaning of these negative events. If I experience a painful breakup and interpret it as evidence I'm unlovable and will never find anyone to marry, that is, for obvious reasons, a less adaptive response than if I interpret it as evidence that relationships can be complicated and sometimes, despite everyone's best intentions, they simply don't work out. Just as mindset interventions aim to change students' attribution styles from statements like *I performed poorly because I'm dumb* to *I performed poorly because I didn't study hard enough, but I'm definitely capable of doing better next time*, CBT seeks to instill in people healthier ways of explaining and interpreting negative events they experience. And there is fairly solid evidence this approach often ameliorates the symptoms of people suffering from anxiety or depression.[50]

So yes, if you are mired in the belief that you are an utterly worthless, stupid person who nothing positive is ever going to happen to, this could make life more difficult for you than it needs to be, and there's a pretty good chance that with the right therapeutic approach (and sometimes medication) you can improve your outlook and feel better as a result. But that's very different from saying that the *root* of all sorts of different societal problems is self-esteem and that broad-based programs to improve it, rather than one-on-one therapies targeting people with very specific mental-health problems, will improve various outcomes. The issue, as with so many half-baked behavioral science ideas, is the jump from claims that are empirically defensible but complex and context dependent to claims that are scientifically questionable but sexy and exciting—and simple.

THE RESIDUE OF THE SELF-ESTEEM CRAZE still lingers. In California and elsewhere, you're less likely to see some of the more over-the-top interventions from the 1990s because most classrooms have moved on, but the fad was sufficiently strong that it left an imprint in state law: during the craze, many states passed legislation explicitly referencing self-esteem, mostly in the area of education.[51] Still on the books in my home state of Massachusetts, for example, is a statute dictating that "each public school classroom provides the conditions for all pupils to engage fully in learning as an inherently meaningful and enjoyable activity without threats to their sense of security or self-esteem."[52] Other statutes codified into law the idea that self-esteem boosting should be a classroom priority, though of course the actual situation on the ground varies from place to place.

As for Vasconcellos, he never lost his interest in self-esteem, even if the publication of *Toward a State of Esteem* marked the triumphant end of his group's three-year mission. His advocacy continued to shape some of the state's legislative priorities. In 2001, he pushed a bill under which, as the *San Francisco Chronicle* explained, "probation officers would get additional training in how to identify low self-esteem among juvenile offenders—and fix the problem through the encouragement of 'personal security, selfhood, affiliation, mission and accomplishment.'" Buried very deep in the story is a frustrated quotation from a state corrections official: "We think it establishes a bad precedent by replacing training based upon scientific research with training based on somebody's opinion."[53] The bill was vetoed by the then governor, Gray Davis.[54]

Vasconcellos's final term, during which he chaired the state senate's Committee on Education, nicely epitomized the mix of pragmatic deal making and New Age idealism that defined his career. "In his first year [of that term] he enacted a program which was signed into law that provided $200million [*sic*] annually for California students in the lowest performing schools," notes a state guide to his archive papers.[55] But in his very final year in office, he was unable to pass Training Wheels for Citizenship, a law that would have granted California minors ages fourteen and up fractional votes in state elections, and that comes across as, well, very Vasco.[56]

Setting aside other aspects of his considerable and complicated legacy, it's hard to deny that John Vasconcellos succeeded in his goal of enhancing the societal importance of self-esteem—a trajectory that took him from being a nationwide target of mockery to being the recipient of praise by a future president. The self-esteem movement's mix of weak science and wide impact became a template for interventions to come.

# 2

# THE SUPERPREDATORS AMONG US

By 1995, America was in the midst of a deepening panic over violent youth crime. Children were both killing and being killed at an alarming rate. From 1985 to 1993 the rate of juvenile homicide arrests had almost tripled,[1] while the murder rate for fourteen- to seventeen-year-old males had jumped from 7.0 to 14.4 per 100,000 members of that cohort during that span.[2]

The high overall murder rate in the United States wasn't new—it had jumped in the mid-1960s and never come all the way back down[3]—but this explosion of youth violence was. News consumers, as a result, were buffeted by story after story about bafflingly tragic murders involving kids. There was a sense in the air that something profoundly evil had gripped the nation's youth; headlines like "Teen-Age Gangs Are Inflicting Lethal Violence on Small Cities" were common.

Stories of young people committing unspeakably violent acts, sometimes gang related and sometimes seemingly random, were everywhere. In Washington, D.C., in 1993, for example, a seventeen-year-old received two life sentences for a firebomb attack on an apartment that killed a one-year-old; the attack was a group effort that "police [said] was a botched attempt to silence a woman who saw one of the five teenagers rob and shoot a cabdriver."[4] And in February 1992, five teenagers in New Jersey, the youngest just fourteen, garroted seventeen-year-old Robert A. Solimine Jr., but not before demanding that he recite the Hail Mary prayer. According to police, "The plan evolved over several months during weekly meetings that mixed elements of religion and Hollywood gangster movies."[5] The era produced some truly jaw-dropping reports, such as this one from *The New York Times* in 1994: "Despite his brother's frantic effort to save him, a 5-year-old boy was thrown to his death from the 14th-floor window of a public housing building on Thursday night by two older boys for whom he had refused to steal candy."[6]

Few cases better dramatize the horrors of this period than that of Robert "Yummy" Sandifer. On an "unusually chilly" September night in 1994, Yummy, just eleven years old and four foot six, was shot in the back of the head by a member of the Black Disciples gang in Chicago.[7] As with so many of the high-profile murders of this period, the killers were young: Cragg Hardaway, arrested for the murder, was sixteen at the time, and his brother Derrick, arrested for driving the getaway car, was just fourteen.

Any murder of an eleven-year-old is a tragedy, but what made the case of Yummy national news was that he himself had been implicated in a killing just two weeks earlier: Yummy was a member of the Disciples, too, and had recently been instructed by older members to attack a rival gang. He followed the order. Firing into two groups of young people on the Far South Side, he accidentally killed Shavon Dean, a fourteen-year-old bystander, and wounded two others. "Police descended on the gang," wrote Nancy Gibbs in *Time*, and as a result "Yummy became a liability" to the Black Disciples.[8] A police manhunt for the boy ensued, but three days later the authorities discovered that his fellow gang members had gotten to him first; the search ended "in a dank, graffiti-scarred pedestrian underpass, where the boy was found dead in a pool of blood and mud," as a *New York Times* reporter in Chicago put it.[9]

Yummy never really had a chance, explained Gibbs in her cover story, which was accompanied by a photograph of the boy looking hard—as hard as a child that young can look, at least—above the words "SO YOUNG TO KILL / SO YOUNG TO DIE." His mother was an addict, his father was in jail by the time Yummy was murdered, and Yummy had had countless run-ins with a legal system that couldn't seriously punish his transgressions because of his status as a minor. "As a baby he was burned and beaten," Gibbs wrote. "As a student he often missed more days of school than he attended. As a ripening thug he shuttled between homes and detention centers and the safe houses maintained by his gang." A local grocer described Yummy in language not usually leveled at eleven-year-olds: "He was a crooked son of a bitch." It was clear that society had failed Yummy at every level. "What you've got here is a kid who was made and turned into a sociopath by the time he was three years old," said the Cook County public guardian Patrick Murphy.

It was bad enough that shockingly young kids were murdering one another, but what helped ramp youth crime fears into overdrive during the 1990s was the possibility that this violence would breach racial boundaries and threaten white people, who had, to that point, been largely insulated from the increase in youth crime. As is always the case in America, during this period whites were less likely to be the victims of violence than blacks; in 1994, for example, black people were 20 percent more likely to be the victim of a violent crime than whites.[10]

These racial fears spread largely through anecdotes and rumors that often mixed real events with feverish and colorful speculation. Take the panic over "wilding," for example. It sprang forth from another crime that made national news: the April 19, 1989, case of a female investment banker who was jogging in Central Park and attacked and left for dead by some of the members of a "loosely organized gang of 32 schoolboys whose random, motiveless assaults terrorized at least eight other people over nearly two hours," as the *Times*'s initial reporting put it.[11] The most notorious fallout of the crime was the arrest and subsequent conviction of the so-called Central Park Five, five innocent black and Latino teens ages fourteen to sixteen at the time who were coerced into confessing and, years later, released and awarded significant damages in a civil settlement with the City of New York after the real perpetrator was apprehended. (Donald Trump famously took out a full-page newspaper advertisement calling for their execution in four New York dailies,[12] and shortly before the 2016 presidential election, remarkably, reiterated that he thought they were guilty; by then they had been free for years.)

During its investigation, the NYPD brought in many youths for questioning, and the term "wilding" originated from some of them, who apparently explained their random attacks by saying, as one detective put it, "We were going wilding."[13] Soon, the term exploded across newspaper pages all over the country. "Wilding" denoted the idea of a group of kids—black kids, if the racial tenor of the coverage is any indication—randomly attacking passersby. It quickly became a fixation of the New York papers; as a trio of Rutgers researchers would later report, from 1989 to 1997 the four major New York dailies published 156 articles referencing the concept.[14] This even though the term was ill-defined. It was used to reference everything from rape and assault to the yelling of taunts, and some of the kids who were supposedly doing the wilding, or who supposedly had friends or classmates who were, told reporters they had never even heard the term.[15] But it didn't matter that the term was vague and that no one showed, convincingly, that it was a significant problem deserving of so much coverage; the idea of whites being attacked by out-of-control hordes of black youths was enough for it to stick in the public imagination.

SO IT'S CLEAR THAT BY 1995 the nation was primed to be receptive to spectacular claims about youth crime—particularly racially tinged ones. It was then that John J. DiIulio, a young political scientist at Princeton University, stepped forward and introduced America to the concept of the superpredator.

DiIulio fashioned himself a tough, no-nonsense truth teller at a time of increasing—and, in his opinion, dangerous—political correctness. By the mid-1990s it was a widely held belief among criminologists, activists, and others that the justice system was biased against darker-skinned people and too tough on them. But in DiIulio's view, the people who harped on such complaints were suckers who didn't understand just how dire America's crime problem was. "DiIulio disagrees [with the reformers] on every count," wrote James Traub in a 1996 profile in *The New Yorker.* "He believes that the system is not punitive *enough*."[16]

DiIulio had a fairly unusual biography for a Princeton social scientist, and he used his background to market himself as something of an academic outsider: raised in a rough part of South Philadelphia by a father who was in law enforcement and a mother who worked in a department store,[17] he clawed his way to a BA at the University of Pennsylvania, a PhD at Harvard, and, at just thirty years old, tenure at Princeton University. He earned early-career plaudits for his 1987 book, *Governing Prisons: A Comparative Study of Correctional Management,* which was, in part, a call for better and more humane prison management.[18]

But DiIulio would come to be known more for his superpredator concept than for any of his other ideas. He told anyone who would listen that because demographic tides would soon bring a disproportionate number of adolescents, America was about to be invaded by an army of amoral youths who were little more than walking killing machines. As shocking as the idea sounded, it rested on two fairly reasonable-seeming assumptions: The first is that young people's home environments affect their propensity for crime. All else being equal, the more stable and nurturing the environment you are raised in, the less likely you are to commit crimes. Family income is also a powerful influence. "Youth who grow up in families or communities with limited resources are at a higher risk of offending than those who are raised under more privileged circumstances," explains a 2014 report from the National Center for Juvenile Justice. "Those who are very poor or chronically poor seem to be at an increased risk of serious delinquency. The timing of exposure to poverty is especially important" as well; the younger a child is when he or she is living in poverty, the more likely he or she will engage in adolescent misbehavior.[19]

This, of course, doesn't mean rich kids don't commit crimes, or that anyone who commits a crime was raised in a tumultuous home. Plus, low-income people, including kids, are often policed more harshly than wealthier ones, which can partially account for some statistical disparities. But as a matter of general correlation, here common sense and research line up. Poorer kids are more likely to get caught in the juvenile justice system than richer ones for a variety of reasons; these reasons are not fully understood, but likely range from the amount of time they spend unsupervised to the availability of help with behavioral problems to the likelihood they have easy access to unsecured guns or have drugs in their home or immediate neighborhood.

The second reasonably well-founded assumption is that, holding everything else constant, the higher the proportion of young males in a population, the more crime and disorder there will be. That's simply because young males seem to be more likely to commit crimes than any other group (at the individual level, simply getting older makes someone less likely to offend; criminologists frequently talk about criminals "aging out" of their prime crime-committing years, though there's at least some debate about how generalizable this claim is to non-American contexts[20]). The actual connection between a country's youth population and its level of crime is, as it turns out, rather complicated, and this simple correlation doesn't always hold. But the general concept of a "youth bulge" causing societal trouble is taken seriously enough by experts that CIA analysts, for instance, have examined this factor when trying to understand political instability abroad.[21]

So neither of these ideas—that home-life environmental factors are correlated with criminality or that a youth bulge could affect the overall crime rate—was, or is, all that controversial in a vacuum. And DiIulio, again, was not some frothing demagogue but rather an established, respected young political science professor with the Princeton name behind him. He wasn't seen as a radical right-winger. Rather, he was a Democrat with fairly heterodox politics; he opposed Bill Clinton's cuts to welfare and would later serve as George W. Bush's czar for faith-based initiatives.[22]

DiIulio introduced the term "superpredator" to lay readers in a 1995 *Weekly Standard* article headlined "The Coming of the Super-Predators." Superpredators, DiIulio wrote, were a class of criminal whose members "kill or maim on impulse, without any intelligible motive." In one important way, he wrote, they were dissimilar to other human beings: "They live entirely in and for the present moment; they quite literally have no concept of the future. As several researchers have found, ask a group of today's young big-city murderers for their thoughts about 'the future,' and many of them will ask you for an explanation of the question." As a result of this inability to heed any imperative other than immediate impulse, superpredators "place zero value on the lives of their victims, whom they reflexively dehumanize."[23]

What was creating these superpredators? Here and elsewhere, DiIulio leaned heavily on the concept of "moral poverty"—the poverty, as he put it in his *Weekly Standard* article, of "being without loving, capable, responsible adults who teach you right from wrong. It is the poverty of being without parents and other authorities who habituate you to feel joy at others' joy, pain at others' pain, happiness when you do right, remorse when you do wrong. It is the poverty of growing up in the virtual absence of people who teach morality by their own everyday example and who insist that you follow suit." "In the extreme," DiIulio continued, "moral poverty is the poverty of growing up surrounded by deviant, delinquent, and criminal adults in abusive, violence-ridden, fatherless, Godless, and jobless settings." Dysfunctional families and neighborhoods and cities give rise to moral poverty, in this view, and moral poverty, in turn, births superpredators.

DiIulio developed his theory to explain a particular rise in crime at a particular moment, but in doing so, he was building upon and borrowing from the idea of the "underclass"—a concept that had, by then, become a fixture of the national conversation about urban decay. The underclass was understood as a particularly troubled subset of the poor who were *truly* cut off from mainstream society and therefore as distinct from the broader population of impoverished Americans.

There was endless debate over where the boundaries of the underclass concept should be drawn. A useful 1991 article in the *Annual Review of Sociology* by the sociologist Carole Marks lays out some of the competing definitions that had been employed to that point by various thinkers. Ken Auletta suggested the underclass should consist of four subtypes: "hostile street and career criminals, skilled entrepreneurs of the underground economy, passive victims of government support and the severely traumatized." Nicholas Lemann argued the term should simply include those immersed in "poverty, crime, poor education, dependency, and teenage out-of-wedlock childbearing." "[The social scientist Christopher] Jencks is more to the point," writes Marks, "noting 'by common consensus [the underclass] includes only the undeserving poor.'"[24] So that's the key point here, definitional quibbling aside: Those who lack resources but are hardworking, rule following, and unthreatening—respectable, in general—are poor, but members of the underclass lack these redeeming qualities. They are a more troubled lot.

But troubled why? That was, and is, a source of much disagreement. In her article, Marks summarizes the simmering debate between "culturalist" accounts, which blame the underclass's plight on their own cultural deficits (unwillingness to work hard, a lack of sexual impulse control, and so on), and "structuralist" accounts, which pin the blame on economic structures and changes beyond the underclass's control. She points out that it is an oversimplification to treat this as an either-or, and in her review she expresses hope that sociologists and other theorists will move beyond this dichotomy. Toward the end she cites the work of the renowned sociologist William Julius Wilson, who posited the "notion of cultural pathologies with structural origins." While this view "has so far been met with heavy skepticism[, i]t may, however, signal a first step in this process." (An oversimplified example can show how intuitive this idea of hybridization is: If a white fifteen-year-old from a rich neighborhood has a good academic work ethic and a black fifteen-year-old from a poor neighborhood doesn't, one obvious reason might be that the former, but not the latter, lives in an environment replete with examples of how working hard can lead to high incomes and the other perquisites of societal success, and can apply his academic motivation in a well-resourced school setting with engaged teachers. The structures of their environments contribute to the habits they develop, in other words, which in turn feed back into their prospects for future success. In this case, it wouldn't be accurate to say that either individual behavior or structural factors tell the *entire* story, even if the latter has a tremendous impact on shaping the former.)

Suffice it to say this fight is still ongoing today, but for our purposes what matters most is that DiIulio leaned heavily toward culturalist explanations of the underclass and adopted many of the culturalists' arguments for his superpredator concept. His main innovation, if you could call it that, was an even more pessimistic assessment of a downtrodden group's behavior and prospects. Staunchly conservative culturalists were of course unflinching in their critiques of underclass behavior, but even they weren't generally making DiIulio's claims of total detachment from human empathy and reason. Superpredators, then, could be seen as the worst of the youth underclass.

Like many other underclass culturalists, DiIulio was careful to note that "kids of whatever race, creed, or color are most likely to become criminally depraved when they are morally deprived," nodding to a deracialized conception of superpredation. DiIulio pointed out the possibility of superpredators popping up in non-inner-city settings, too, albeit with less emphasis than the imminent threat of urban superpredators. "While the trouble will be greatest in black inner-city neighborhoods," he wrote in the *Weekly Standard* article, "other places are also certain to have burgeoning youth-crime problems that will spill over into upscale central-city districts, inner-ring suburbs, and even the rural heartland." He could speak that colorblind language when he wanted to.

But in reality, superpredators, like the underclass, were usually coded as black in the public imagination, and that's partly because DiIulio's focus was clearly on black superpredators; it was young black people, he argued, who were getting hit hardest by the culture of moral poverty. DiIulio took the conservative culturalist stance that black people were primarily being held back not by societal structures or oppression but by their own cultural and behavioral shortcomings, though his thoughts on the matter were sometimes conflicted. In some of his writing on the subject, such as "My Black Crime Problem, and Ours," published in *City Journal* in 1996 (the title is a reference to Norman Podhoretz's famous 1963 essay in *Commentary,* "My Negro Problem—and Ours"), DiIulio dances between empathy for black people over the oppression they have faced and skepticism of certain liberal ideas about discrimination in the criminal justice system. "I find almost nothing in the empirical research literature on racial disparities in sentencing to justify their fears and frustrations," he writes of African Americans who complain of unfair treatment. "But that does not mean their fears and frustrations are without any sort of empirical basis, including the kind that, for most of us, counts more than any other—our own lived experiences and those of our family and friends."[25]

To hammer home that idea, DiIulio cites an instance from his youth in which his work buddy at a pizza joint, an illiterate then-forty-something black man, was pushed up against the wall by police with no explanation apart from "'Sorry, kid, we got a call that somebody was stealing.' But the only 'somebody' they grabbed was the black man, and the only apology they issued was to the white boy." He explains that when interracial fights broke out during his youth, "everyone ran when the cops came. But we white boys knew that if we got caught and showed due deference to the officers (*never* mouth off to a panting cop), we'd get lectured, get our hair pulled, or (at most) get our fathers called. We also knew the black boys would get that and worse—slapped, clubbed, and maybe arrested." DiIulio acknowledges, in short, that the system has been stacked against black people. But "the bottom line of most of the best research is that America's justice system is not racist, not anymore, not as it undoubtedly was only a generation ago."

So DiIulio, like many modern racial conservatives before and after him, and like the culturalists participating in the underclass debate, argued that there was nothing *inherently* inferior about African Americans. He believed that if they embraced different values, they could be just as successful as whites, because the structural obstacles had already been overcome. Yet he never quite explained how the structural problems, particularly racist law enforcement, that he himself acknowledged were serious in his own youth could possibly have been solved so quickly.

If the *Weekly Standard* article was DiIulio's superpredator hors d'oeuvre, the main course came in a 1996 book DiIulio co-authored with William J. Bennett, who had been Ronald Reagan's education secretary and George H. W. Bush's crusading antidrug czar, and John Walters, then the executive director of the Council on Crime in America. *Body Count: Moral Poverty . . . and How to Win America's War Against Crime and Drugs* consisted mostly of a mix of scary stories about youthful offenders and the debunking of what the trio viewed as liberal (and some conservative) canards about crime and the justice system. Most important, though, the book echoed DiIulio's *Weekly Standard* argument that America was about to be beset by a wave of superpredators. Key to the superpredator meme was the idea that whatever was going on was genuinely *new*; the violence splattering itself across the streets of America represented not some blip or upward tick that could reverse itself but a jarring and potentially permanent change.

Indeed, the authors of *Body Count* argued, things were going to get much, much worse. The coming carnage was foretold by simple demographics, by a "ticking crime bomb"—variations of this phrase would appear in superpredator coverage over and over, in magazines and newspapers—for which the evidence was "overwhelming."[26] The argument was simple: A certain proportion of young men growing up mired in moral poverty became superpredators, and there were going to be more and more young men growing up mired in moral poverty in the coming decade. Therefore, things were about to get apocalyptically violent.

DiIulio had some big-name help in spreading these scary ideas.

Most important was James Q. Wilson, his dissertation adviser at Harvard, who was then at UCLA (and is since deceased). Wilson was a political conservative who is to this day seen as one of the most important criminologists, if not social scientists, of all time; he has been frequently lauded by social scientists from across the political spectrum as a careful, rigorous thinker and writer (though his legacy is certainly complicated by his role in developing the theory of "broken windows" policing, which many social scientists believe contributed to the over-incarceration of low-income and minority men for committing "quality of life" violations). By the time the superpredator concept arose, Wilson's articles and books (with titles like *Varieties of Police Behavior* and *The Investigators: Managing FBI and Narcotics Agents*) had deeply influenced a generation of criminologists and other social scientists. The fact that Wilson helped promote the superpredator meme—and that even when he wasn't doing so directly, his name was frequently connected to it—likely aided its spread a great deal (it's also worth pointing out that by 1999, Wilson had said he was likely wrong to have done so[27]).

Wilson also had James Alan Fox, then and now a very media-savvy Northeastern University criminologist, on his side. DiIulio, Fox, and Wilson were convinced that the demographic and crime-stats evidence pointed to a coming bloodbath, and the media eagerly propagated this story line being offered by Ivy League–credentialed experts, often seeming to revel in language that painted young killers as subhuman. " 'Superpredators' Arrive: Should We Cage the New Breed of Vicious Kids?" asked *Newsweek*.[28] When violent teens weren't subhuman, they were incendiary, explosive; the superpredators represented a "teenage time bomb," as *Time* put it in a cover story of its own.[29]

So what was to be done? In both the *Weekly Standard* call to arms and other works, including *Body Count*, DiIulio and his colleagues expressed deep skepticism of structuralist approaches to fighting crime.

"'Body Count' has virtually nothing to say about jobs, housing, or health care," noted James Traub in his *New Yorker* profile. It did, however, have plenty to say about the importance of incarceration and religion, and DiIulio often echoed these themes when he wrote or spoke about superpredators. He concluded his *Weekly Standard* article, for example, with a simple, stark motto: "'Build churches, not jails'—or we will reap the whirlwind of our own moral bankruptcy."

But researchers who are receptive to the idea of a causal relationship between church attendance and positive outcomes among at-risk youth have a more nuanced view than this: In one 1986 book chapter DiIulio cites, the economist Richard Freeman writes that "at least some part of the churchgoing effect [that he discovered] is the result of an actual causal impact,"[30] but then he exhibits significant caution about not overstating or oversimplifying that result. Later on, DiIulio himself, summing up other work, would write for *The Brookings Review* (the Institute's quarterly magazine), that "the potential of church-going and other religious influences to improve the life prospects of poor black urban youth is in part a function of how church-going and other faith factors influence how young people spend their time, the extent of their engagement in positive structured activities, and the degree to which they are supported by responsible adults."[31] If this is true, it would suggest that DiIulio's rejection of "root cause" approaches in favor of religion was misguided: That is, if what matters is access to structured activities with adult supervision and poor urban areas lack non-church-based options in this department, then why not fund a *variety* of such programs, both religious and non?

In any case, the belief in superpredators led to an obvious policy prescription: getting tougher on violent youth crime. In his *Weekly Standard* article, DiIulio bragged that "no one in academia is a bigger fan of incarceration than I am." By his estimate, he said, America would "probably need to incarcerate at least 150,000 juvenile criminals in the years just ahead . . . We will have little choice but to pursue genuine get-tough law-enforcement strategies against the super-predators. But some of these children are now still in diapers, and they can be saved."

AS IT TURNED OUT, DiIulio and his tough-talking colleagues were wrong on just about every count. America had as much reason to worry about "superpredators" as it did to worry about other 1990s bugaboos like killer bees and the Y2K bug. When one looks at violent youth crime graphs since the time DiIulio and his ilk were making their scariest prognostications, the trajectory is not a terrifyingly steep uphill climb but rather a gentle decline that hasn't really ticked back upward since.[32]

No one knows exactly why violent youth crime has fallen the way it has. Nor is there agreement over why violent crime in general started to decline in the early 1990s. Theories range from the dissolution of the crack trade to the introduction of restrictions on lead paint (which can cause impulsive and violent behavior) to the relatively strong economy and job market of the 1990s to, perhaps most controversially, the long-term results of easier access to abortion (fewer unwanted kids means fewer kids at risk of committing violent acts).

As for why youth crime went up in the first place in the late 1980s and into the 1990s, most criminologists who have studied the issue closely believe that a great deal of the rise can be attributed to easy access to guns, largely as a result of arms races associated with gangs fighting over the crack trade.[33] As that business grew increasingly cutthroat, even low-level dealers felt they had to carry guns to protect themselves. And more guns in teenagers' hands is an inevitable recipe for bloodshed. Contra DiIulio's most striking claims, the research available on violent youth crime in this period doesn't really support the idea that it was "unintelligible" in the manner he suggested. It was reckless and tragic and adolescent and foolish, but it wasn't, in most cases, teenagers randomly attacking others because of a deep moral bankruptcy; rather, these killings tended to occur in the context of drug violence and its terrible cascading effects.

Whatever the cause of the subsequent drop, a look back at the specific claims DiIulio and his fellow travelers made, and how they made them, reveals some striking shoddiness; these were never seriously rendered, compelling arguments. To take one key example, DiIulio and Wilson both based some of their dire predictions on the idea that 6 percent of young people, nationwide, were likely to turn into serious violent criminals. This idea originated from the so-called Philadelphia birth cohort study, which tracked almost ten thousand boys born in 1945 from the time that they were ten until eighteen years of age and recorded their instances of police contact. In making their claims, the superpredator evangelists simply assumed that any boy in the study who had had five or more instances of police contact was a serious criminal.

The 6 percent figure, they argued, could be used to project approximate future rates of violent crime. As Wilson wrote in the 1995 book *Crime*, which he co-edited with the criminologist Joan Petersilia, "By the end of this decade there will be a million more people between the ages of fourteen and seventeen than there are now . . . This extra million will be half male. Six percent of them will become high rate, repeat offenders—30,000 more young muggers, killers, and thieves than we have now. Get ready."[34] That prediction also found its way into the *Weekly Standard* article. It was on the basis of the 6 percent number, DiIulio wrote, "that James Q. Wilson and other leading crime doctors can predict with confidence that the additional 500,000 boys who will be 14 to 17 years old in the year 2000 will mean at least 30,000 more murderers, rapists, and muggers on the streets than we have today."

Setting aside the challenge of extrapolating from one city to the entire country, the problem is the original study didn't, in fact, find that 6 percent of male Philadelphia youth were serious offenders. Wilson was simply wrong to assume that just because a kid in the study came in contact with the police five or more times, he was a criminal. As the gang researcher James C. Howell later put it, "In fact, only one-third of the police contacts resulted in an arrest, and only half of this group's police contacts resulted in a court adjudication of delinquency."[35] That is, the percentage of the cohort found guilty of *any* form of delinquency was far lower than 6, and plenty of those kids were convicted of crimes that weren't violent at all; a 1995 reanalysis of the data found that "although chronic offenders [in the study] do commit more petty crimes than others, they are involved in a 'cafeteria style' delinquency and are not as dangerous as has been generally believed."[36] In short, DiIulio, Wilson, and their colleagues seriously overestimated the rate at which members of the Philadelphia birth cohort study committed serious crimes, and then extrapolated this misunderstanding into a scary prediction.

There were other, more fundamental problems with the superpredator concept, too. For one thing, DiIulio and his colleagues never even fully defined what a superpredator was, except, in essence, a young person who does really bad things and who doesn't appear to exhibit remorse. This is a common phenomenon when half-baked ideas go viral; these ideas are often characterized by a certain *conceptual fuzziness*. That is, the idea in question, upon closer inspection, seems to lack clear boundaries and definition. It's often the case that the idea feels right, especially to nonexperts, and makes enough intuitive sense to spread—especially among lay audiences or those who find the idea ideologically useful—but that if you poke and prod it a little bit, you realize that it isn't, well, baked all the way through. When handled, pieces break off, or the idea crumbles entirely.

In an academic or public policy setting, a term like "superpredator," if it is to be useful, should be defined carefully and specifically. Any criminologist will tell you that otherwise normal people do horrible things all the time and that adolescents, in particular, can be shockingly reckless and shortsighted. So when one of them does something shockingly reckless, how can you tell whether he is a superpredator or just, well, a teenager? If a fourteen-year-old raised in a broken home and involved in the crack trade kills someone, exhibits no remorse in the moment, but deeply regrets what he did when he is a little bit further along developmentally—in exactly the same way so many adults regret and are baffled by something *they* did as adolescents—is it fair, or criminologically useful, to brand him a superpredator who is ostensibly incapable of normal human emotions? Why should he be treated differently, morally speaking, from, say, a child soldier? What differentiates a superpredator from the many other people who commit terrible acts? How is one diagnosed? Do they respond differently to threats of punishment or social sanction from loved ones? DiIulio didn't bother seriously investigating any of these key questions.

The concept of moral luck, introduced by the philosopher Bernard Williams and popularized by Thomas Nagel, calls attention to how the role of luck is elided in many of our moral judgments, and also helps illuminate certain problems with the superpredator idea. Perhaps most relevant here is what Nagel calls "circumstantial" moral luck. As he points out in *Mortal Questions*, everyday German citizens faced a profound moral test after the Nazis took over. While "most of them are culpable for having failed this test . . . it is a test to which the citizens of other countries were not subjected, with the result that even if they, or some of them, would have behaved as badly as the Germans in like circumstances, they simply did not and therefore are not similarly culpable."[37] Luck, and luck alone, partially explains the difference between being a Nazi collaborator and living life without committing such an atrociously bad moral act. Like so much else, the opportunity to live a moral life is distributed unequally, and whatever harm he has done, an armed thirteen-year-old involved in the crack trade clearly didn't have the same chance to do good as many others had. How would a truly thought-through theory of superpredators deal with this reality and apply it to questions of punishment and proportion?

To elide such questions is to engage in conceptual fuzziness, and DiIulio and his colleagues' theorizing about superpredators is rife with such fuzziness. But it didn't matter: the concept of a superpredator was scary and evocative enough that the public—and, crucially, lawmakers—glommed on to it. The idea took on a very loud and influential life of its own despite its obvious-in-retrospect shortcomings. And crucially, the chief proponents of the superpredator concept were able to mostly sidestep any real stress testing among bona fide experts. In fact, there's almost no discussion of superpredators in peer-reviewed academic journals, except in the form of later debunking and cultural analysis of the spread of the term, which is remarkable given how many people were deeply convinced that the nation was about to be overrun with members of this scary group of sub-underclass denizens.

In his very insightful book *The Ideas Industry: How Pessimists, Partisans, and Plutocrats Are Transforming the Marketplace of Ideas*, the Tufts University international politics professor Daniel Drezner posits two general modes in which academics and others present ideas to the public.[38] "Thought leaders" are very confident, not particularly analytical or critical, and tend to focus on their "one big idea" that they are convinced can change the world. "Public intellectuals," on the other hand, see things in a somewhat more nuanced, complex light; they're more likely to critique ideas they see as lacking, and are generally skeptical of the framework of "This one idea can explain the world."

Though it is, by design, a simplified account that leaves out certain middle ground, Drezner's model is illuminating. And it can be usefully extended to situations in which someone engages in code switching between the thought leader and the public intellectual modes of discourse. DiIulio's academic work was, by many accounts, generally careful and rigorous, while his public proclamations about superpredators were much less so. Academic peer review is by no means perfect (as we will see), but had he attempted to publish peer-reviewed research about superpredators, he would have potentially been forced to answer questions that were less likely to be posed by lay audiences: what, exactly, the term meant, whether his use of it meshed well with previous theorizing and empirical work on violent youth offenders, whether the demographic projections he was relying upon were sturdy, and so on. He would have had to define his terminology thoroughly and coherently. But he instead appears to have focused mostly on mass-market writing and to have spoken primarily to policy makers and politicians, few of whom were equipped with the tools to ask tough, skeptical questions about the concept's coherence but many of whom were inclined, or politically motivated, to buy what he was selling. DiIulio figured out how to speak the language of the mid-1990s anticrime thought leader—rather than public intellectual—and he was rewarded for it.

And because of a lack of skepticism among the public and politicians, the superpredator concept, despite its thinness, had an important and pernicious effect on the national debate over youth crime. Tough-on-crime lawmakers quickly seized on it to further ramp up their preferred policies—incarceration, incarceration, incarceration—and to gin up fears of an impending wave of even worse youth crime. "Legislators were not swayed by scholarly evidence" as the crime panic unfurled itself, writes Ashley Nellis, a senior analyst at the Sentencing Project, in her book *A Return to Justice: Rethinking Our Approach to Juveniles in the System*. "As a result, terms like 'superpredator' made their way into state legislation as if they described an actual phenomenon rather than an invented myth." Nellis goes on to cite several examples, such as the New Mexico senator Pete Domenici introducing a 1996 law that he said was designed to "update laws to deal with the 'superpredator,' the increasingly violent juvenile criminal," and Bob Dole using the term—and referencing scary demographic projections—during his 1996 presidential bid.[39]

In Nellis's telling, though, it was probably Bill McCollum, a Republican representative from Florida's Eighth District and chair of the House Subcommittee on Crime, who most enthusiastically seized on the superpredator meme. She quotes some of his crusading testimony before the House Committee on Early Childhood, Youth, and Families in 1996:

> In recent years, overall crime rates have seen a modest decrease . . . Nevertheless, this general decline masks an unprecedented surge of youth violence that has only begun to gather momentum. Today's drop in crime is only the calm before the coming storm . . . Now here is the really bad news: This nation will soon have more teenagers than it has had in decades. In the final years of this decade and throughout the next, America will experience an "echo boom"—a population surge made up of the children of today's aging baby boomers. Today's enormous cohort of five-year-olds will be tomorrow's teenagers. This is ominous news, given that more violent crime is committed by older juveniles (those 15–19 years of age) than by any other age group. More of these youths will come from fatherless homes than ever before, at the same time that youth drug use is taking a sharp turn for the worse. Put these demographic facts together and brace yourself for the coming generation of "super-predators."[40]

By no means did Republicans have a monopoly on such talk. Few politicians, regardless of party, could resist the tides of the tough-on-crime 1990s—a period during which many Democratic lawmakers were staking out more centrist and conservative stances on issues like crime and welfare anyway. So DiIulio's pseudo-academic ideas found their way into Democratic lawmakers' offices as well. "Kids nationwide are committing murders and other violent felonies at ever-increasing rates, and the trend among New York City teenagers is by far the most frightening, according to a congressional study," noted a New York *Daily News* dispatch from 1996. "Juveniles ages 13 to 17 are more likely to commit violent crimes than any other age group, and the projected growth of the 'at risk' 13-to-17 age group over the next 10 years will mean that 'the nation will see a dramatic increase in crime.'" The congressional office in question? Charles Schumer's.[41] Alarmed by the supposed demographic trends, Schumer "proposed changes in the Juvenile Records Act to unseal teen crime records and allow judges and cops to learn of previous violent crimes." Then First Lady Hillary Clinton mentioned superpredators during a call to "take back our streets from crime, gangs, and drugs" in a 1996 speech broadcast on C-SPAN[42]—a reference that later drew the ire of civil rights groups, especially during the 2016 presidential election. And while he didn't use that exact term, Joe Biden did refer in a 1993 speech to "predators," a "cadre of young people, tens of thousands of them, born out of wedlock, without parents, without supervision, without any structure, without any conscience developing because they literally . . . because they literally have not been socialized, they literally have not had an opportunity."[43]

Now, the question of *exactly* what effect the superpredator idea had on lawmaking is complicated to answer, because by the time it went viral American states had already spent several years tripping over one another in a race to get tougher on youth crime by, for example, making it easier to charge children as adults. But many experts believe in a direct link. "Based on that (as it turned out) fallacious bit of science, dozens of states passed laws allowing juveniles to be tried and sentenced as adults, with predictably disastrous results," wrote the late criminologist Mark Kleiman in a 2016 *Washington Monthly* essay.[44] And as Human Rights Watch put it in one report from 1999, "The modern-day legend of the coming superpredators has driven legislative approaches to juvenile and criminal justice, overcoming hard facts and sound research."[45]

Whatever its precise, tough-to-pin-down effects on actual lawmaking, the superpredator meme certainly had an impact. Consider the important disconnect between official crime statistics and the perceptions of many Americans—a disconnect that might have delayed subsequent reform efforts. In the mid-1990s, at the apex of superpredator fearmongering, signs of a crime drop had already begun showing up in the data (some in the superpredator crowd simply dismissed this as a temporary statistical blip before things got truly out of hand), but that fact wasn't reflected in public opinion. People were still very scared of crime and convinced it was on the upswing. In one 1996 poll cited by Nellis, for example, "84 percent of respondents thought that violent crime had risen in the previous year, but in reality it had dropped," and she cites another poll "conducted in the early 2000s [that] showed that even though the arrest rates for youth homicide had plunged 69 percent between 1993 and the end of the decade, 62 percent of the American public believed it was still rising."[46]

It's natural that there is going to be some lag between a given statistical change and the public's perception of that change, but a meme as evocative, emotional, and contagious as the superpredator might have opened a wider-than-normal gap between Americans' intuitions about the trajectory of youth crime and the reality. Nellis notes that "few respondents believed crime was a problem in their neighborhoods, suggesting that their estimates were not based on personal experience."[47] But the superpredator meme rendered the idea of out-of-control, senseless violent crime quite cognitively accessible even to many Americans personally unaffected by these dangers. This is an example of the availability heuristic—"I can easily come up with examples of X, so X must be pretty common"—which is a well-known cognitive quirk that often causes people to develop misconceptions about how likely various events are (compare how easy it is to think of an airliner crash with the statistical probability of one occurring). In the years since, as we will see, some of the legislative excesses of the youth-crime-panic era have been rolled back, but would they have been rolled back sooner if the superpredator craze hadn't taken over? It feels like a reasonable assumption.

One curious fact about the history of this meme is the relative quiet of the experts who could have pushed back against it. The superpredator idea spread so rapidly in part because there wasn't much public debate about the claims of DiIulio, Wilson, and Fox, other than a smattering of skeptical op-eds penned by academics. Franklin Zimring, a professor at the UC Berkeley School of Law and one of the idea's few outspoken contemporaneous critics, wrote in 2013 that while there was plenty of skepticism within the academy among criminologists and demographers, "for the most part . . . the academic reaction to the demographic time bomb rhetoric was silence, whether respectful or not."[48] When we spoke in 2017, he said he partly blamed that silence on the fact that youth criminology wasn't a particularly well-populated field, so there simply weren't all that many would-be debunkers around. "People were otherwise occupied," he said. "It wasn't politically risky or anything, and it wasn't that there was no market, that news media wouldn't have listened. It just didn't happen. This isn't a thick area; if John DiIulio went into the global warming business with statistics like this, it would take milliseconds before an awful lot of PhDs would be marching in with critiques, highly public critiques."

The lack of debunking from experts is only part of the story, though. At root, when society accepts a false story, it's often because that story tells society something it wants to hear, or something that seems to explain what's going on at the moment. DiIulio might have sought to portray the concept of "superpredator" as not being about race per se, but its spread was racially fraught from the start. There's a strong case that the meme proliferated so wildly because it fit a story line many white Americans were prone to believe: *Black crime is fundamentally different from white crime, and needs to be dealt with differently—and more harshly. That's the only way to get us out of this horrifying mess we're in.*

MOST PEOPLE HAVE TROUBLE GRAPPLING with the idea that normal humans can act in truly terrible ways. Yes, there are some criminals who are so disturbed that they can safely be set aside and put into a behavioral category all their own—true psychopaths, for example, or people with schizophrenia who harm others as a result of delusions or hallucinations. You could spend years interviewing these criminals and you'd come away with few insights about their heinous acts applicable to most humans.

But setting aside this tiny minority of criminals, the sad fact about our species is that plenty of its otherwise ordinary members assault and murder and rape. Sometimes it's Hutus killing Tutsis in Rwanda, sometimes it's white suburbanites showing up armed on the front lawns of black homeowners trying to integrate the Detroit suburbs in the 1950s, sometimes Blackwater contractors murdering Iraqi civilians, and sometimes . . . well, insert any example you want from millennia of well-documented bloodshed and grief. There are all sorts of situations in which humans commit awful acts despite seeming to lack any sort of deep underlying pathology or dysfunction.

This is the sort of fact that makes us, hungry as we are for order and meaning and a sense that the universe is a just place, very uncomfortable. Which can help explain why many societies have gone to great, sometimes creatively sadistic lengths to paint a bold, dark line between normal, good people who don't commit crimes and the ostensibly evil, corrupted, haunted souls who do. That there's so much overlap between the two groups tends to be swept under the rug. To maintain this facade of a distinction, criminals need to be sequestered or marked or ostracized or disfigured—a point illustrated graphically in Kafka's story "In the Penal Colony," which describes a grotesque machine that executes convicts by tattooing the law or rule they broke onto their bodies in excruciating—and very public—fashion over the course of twelve hours.

In many cases, the desire to mark and set apart criminals may be more a psychological protection mechanism than an effective anticrime measure. This impulse is connected to a broader tendency in human psychology—what psychologists call the fundamental attribution error. As a general rule, and all else being equal, we treat in-group and out-group members very differently when it comes to explaining their behavior. If we commit a bad act, or someone close to us does, it's often an utterly shocking aberration. There are myriad contextual explanations for it: the person was tired or hungry or had fallen under the influence of a *truly* malevolent actor. We conclude that the act doesn't reveal the doer's essential character. When it comes to the bad acts of *those untrustworthy guys over there,* on the other hand, we are more likely to attribute whatever happened to something essential. *They're* just like that, so when they commit murder it makes sense; they have traditionally been a murdery people, after all.

This principle is neatly explained by the timeless *West Side Story* vaudeville number "Gee, Officer Krupke." The song starts as the titular cop demands that the ringleader of that no-good street gang the Jets give him one reason not to arrest him on the spot. The Jets, it turns out, have *many* reasons on offer, all of them stemming from their insistence that they are the victims of forces beyond their control.

While the song is clearly intended as a lighthearted lampooning of bleeding-heart attitudes toward youth delinquency, it doubles as an illustration of the fundamental attribution error: to Officer Krupke, the hooligans are bad apples who need disciplining. To the hooligans themselves, they're innocent victims—or at least they know how to play that role. The lead Jet begins his appeal by singing, in part: "Our mothers all are junkies / Our fathers all are drunks / Golly Moses, natcherly we're punks!" The song proceeds to lay out a seemingly endless procession of theories as to why the gang's members are so troubled—ranging from a drug-dealing grandma to cross-dressing siblings (remember that this was the 1950s)—but the overall point is that surely it's someone *else's* fault.

Just as we are prone to excuse the in-group, Jets-style, we also tend to essentialize the out-group. Not surprisingly, a major goal of criminology has always been to figure out why people commit crimes, and one form this quest has taken in the past has been a fixation—sometimes an obsession—with identifying a criminal *type*. That is, a category of person who can be identified, before the fact, as likely to engage in crime, particularly serious crime. In addition to its obvious ties to the underclass concept, DiIulio's "superpredator" category can be placed in a long line of theories attempting to establish broad categories of humans who are criminal by nature and who therefore aren't quite like the rest of us.

DiIulio's superpredator is born of environment, not biology, but for a long time it was biological determinism that drove the search for the criminal type. It was assumed that the criminal had certain inborn features that marked him as distinct from normal humans and that were responsible for his behavioral abnormality. No one pursued this idea more aggressively than Cesare Lombroso, the Italian researcher who is viewed as the father of criminology.

Lombroso, an Italian Jew born in 1835, was trained as a physician, and he developed a fascination with the measurable physical differences between humans. A great deal of his research energy went into simply measuring people—both living and dead ones—and he developed a specific interest in the bodies of the insane and the criminally inclined. As the head of multiple insane asylums in the 1860s and 1870s, he had no shortage of access to these sorts of bodies.[49]

Lombroso became especially intrigued by the idea that those with deep-seated criminal inclinations were more animal-like than the rest of us and that this could be revealed by measuring their physical features. By the time the fifth and final version of his touchstone work, *Criminal Man*, was published, he had theorized a rich taxonomy of different criminal types. Some had committed criminal acts as a result of extenuating circumstances, not any deep-seated flaw—as evidenced by their lack of the physical abnormalities that, in the Lombrosian worldview, indicated more essential forms of criminality. "Criminals by passion," for example, had committed an emotionally driven crime and quickly apologized. (Perhaps the nineteenth-century equivalent of a Jet in Italy would have sung, "Officer Krupke, please understand / It's just I got these passions / that get so out of hand.")

*Born criminals*, on the other hand, represented a much more significant and dangerous strain of criminality. As Mary Gibson puts it in her book *Born to Crime: Cesare Lombroso and the Origins of Biological Criminology*, they were "throwbacks on the evolutionary scale, a freakish reappearance in modern European civilization of its brutish past." Because they weren't quite human, they needed to be treated harshly; they didn't respond the way normal, full-blown humans did to external incentives and didn't feel remorse. It shouldn't come as a surprise that Lombroso and his contemporaries believed some races were more evolved than others and that nonwhite people were much more likely to possess "atavistic physical features," as Gibson puts it, and to have less control over their emotions and violent impulses.[50]

Gibson notes that "Lombroso's emphasis on racial difference was neither original nor exceptional in late nineteenth-century Europe, where many biologists, anthropologists, and statisticians were trying to turn race into a science."[51] This fixation on racial differences left a profound mark on criminology, particularly in the United States. As Khalil Gibran Muhammad shows convincingly in *The Condemnation of Blackness*, ever since the first American census containing arrest data broken down by race was published, in 1890, social scientists worked tirelessly to use that data to spin compelling stories—stories that have often been lapped up by white audiences—about the fundamentally inferior, and more delinquent, nature of blacks.

It's impossible to ignore the fact that there are, in fact, racial gaps with regard to violent crime. To oversimplify and leave out other ethnic groups, black people are more likely to commit and to be victimized by violent crimes than white people. The perpetration gap itself is not a myth and has existed in some form for a long time, but it's a politically explosive and frequently misunderstood subject. There's significant evidence that when certain variables are controlled for, the black-white crime gap closes dramatically. One article by a pair of sociologists, for example, looked at trends in rates of violence among whites, blacks, and Latinos since 1990 and found that "consistent with expectations, structural disadvantage is one of the strongest predictors of levels and changes in racial/ethnic violence disparities."[52] Because blacks face so much more structural disadvantage than whites by any conceivable measure, the crime gap is no surprise. The economist Dionissi Aliprantis takes a different tack, finding that "black and white young males are equally likely to engage in violent behavior, conditional on reported *exposure to* violence" (emphasis mine).[53] Given that, as Aliprantis notes, "twenty-six percent of black males in the USA report seeing someone shot at before turning 12," it's no wonder there is a crime gap.

These articles, and others like them, don't answer the racial-crime-gap questions conclusively, and in fact there are serious methodological difficulties to doing so. When I asked Aliprantis in an email whether anyone had done a big, full-blown between-races correlation relating income or family wealth to propensity to commit crime, he pointed out that racial wealth inequality is so rampant in the United States that there simply aren't enough well-off black families to run such an analysis in a robust way.

The main takeaway from this quick tour of the research is that it's possible to talk about the racial crime gap in a rigorous manner, to put the numbers in context. Essential, in fact. But as Muhammad explains in his book, these statistics have often been used to buttress racist ideas, especially as biological racism fell out of favor around the end of the nineteenth century. "By the 1880s the best scientific efforts to prove the physical inferiority of African Americans had fallen short," he writes.[54] This led to a new set of somewhat more sophisticated (in a loose sense) culturally and behaviorally oriented arguments about black inferiority; out was junk science measuring skull shapes and sizes, and in were newer, more "nuanced" claims that blacks weren't fit to be treated as equal citizens, as conveniently proven in innovative-seeming quantitative tables. "By taking a broad statistical view of the field, it will be possible to found our conclusions on much surer ground," wrote Nathaniel Shaler, a white-supremacist paleontologist and geologist in an 1890 article in *The Atlantic*.[55]

It would be unfair to directly compare Shaler's views to DiIulio's. But the latter's superpredator work seemed to take a different road to a similar destination. Compare some of the language between the two men: As Muhammad points out, Shaler believed that "blacks were incapable of controlling their sexual impulses; they were unable to work together for a common purpose; and, most important, they had no power to delay gratification and plan for the future."[56] DiIulio, too, referenced the idea of blacks who had no power to delay gratification, and mentioned "wolf packs" of amoral monsters preying on the innocent. In 1896, Frederick Hoffman, another prominent white supremacist, wrote that "vice and crime are the only formative influences" in some urban neighborhoods inhabited by African Americans.[57] A century later, DiIulio wrote that many blacks in inner cities were "growing up in the virtual absence of people who teach morality by their own everyday example and who insist that you follow suit."

In the years and decades that followed Shaler's proclamations about black inferiority, the supposedly distinct natures of white and black crime would drive countless important policy decisions. White children, but not black children, had access to the earliest reformatories.[58] White immigrant crime, but not black crime, was excused as the by-product of cruel treatment at the hands of society. Yes, there were hardened white racists who considered Irish and Italians not quite human, but that wasn't society's general thrust; rather, there was a sense, especially among Progressive Era reformers, that these populations could be raised up through institutions like settlement houses, the first of which were founded right around the time Shaler was proclaiming his excitement for statistical race science. Blacks, of course, were explicitly excluded from most settlement houses and from many other turn-of-the-century social services, or their neighborhoods simply didn't get the same goods. And the crime numbers offered an easy justification for such neglect.

IN SHORT, CERTAIN CRIMINOLOGICAL THEORIES, whether pseudo-psychological or statistical, may derive less from solid evidence than from a universal human drive to other-ize criminals in combination with the more recent development of racial hierarchies. If one believes that "certain types" are more likely to be criminals of the sort that can't be reasoned with or controlled through normal measures, after all, racial hierarchies offer a useful hint as to who those "certain types" are.

This applies both to formal theories developed by criminologists and to folk theories about crime. It's a tendency that pops up regularly in a variety of public debates over wrongdoing, from the frequently trotted-out trope that there is something unique in Islam that makes the faith resistant to reform and modernization—and its adherents therefore much more likely to be violent terrorists unswayed by terrestrial concerns or incentives—to Donald Trump's famous fixation on crimes committed by undocumented immigrants (which ignores the fact that by many measures, members of this group commit crimes at lower rates than natural-born citizens).

As for John DiIulio, there's little reason to believe he explicitly intended his idea to reinforce racial stereotypes. Indeed, he would later win plaudits from some black leaders for his work as head of George W. Bush's Office of Faith-Based and Community Initiatives. While that office generated a great deal of controversy and angst among many liberals, who saw it as a breach of church-state separation, it was also met with significant enthusiasm by some black religious leaders and their congregants (in part because the office brought with it opportunities for them to receive federal funding).[59] While there, he worked with his friend the much-admired African American minister and activist the Reverend Eugene F. Rivers to address social problems via churches, for instance.[60]

But it would be foolhardy to ignore the reality here: whatever other progress was made in the twentieth century, by the end of it a large chunk of society remained all too willing to accept half-baked statistical proof of an idea that caused many black people caught in the justice system to be treated as irredeemable monsters. At a certain point, however these stories are dressed up, the same message shines through: Black crime and black life are just different. The stats prove it. And it's possible to arrive at this conclusion through an all-nature account (blacks are biologically inferior), an all-nurture account (black culture and values are inferior), or some hybrid. The point is the same. It's little wonder, in light of all this, that few people seemed all that interested in white superpredators, even if DiIulio did nod to their existence, or that there's been a comparative dearth of interest in the white underclass (though to be fair, some right-of-center writers and thinkers, like Charles Murray and J. D. Vance, have employed at least partially culturalist theories to explain white underclass behavior). The vast majority of superpredator discourse concerned black neighborhoods and black people, and it convinced many white Americans the inner cities were teeming with countless not quite fully human violent teenagers, with hordes more on the way.

IN 2012, JOHN DIIULIO, James Alan Fox, and James Q. Wilson signed on to a Supreme Court amicus brief that completely renounced the superpredator idea.[61] The brief, which sat alongside another amicus brief co-authored by Franklin Zimring and other noteworthy superpredator skeptics, was written in support of Evan Miller and Kuntrell Jackson, two black men who were challenging the life sentences—mandatory under the law—they'd received for their involvement in murders committed when they were fourteen (Miller was directly involved in the murder he was charged with, while in Jackson's case he was charged even though it was his friend who pulled the trigger during a botched robbery attempt). The question before the court was whether laws stipulating mandatory life imprisonment for minors constituted cruel and unusual punishment.

The language is straightforward throughout the amicus brief. "The fear of an impending generation of superpredators proved to be unfounded," argue the authors. "Empirical research that has analyzed the increase in violent crime during the early to mid-1990s and its subsequent decline demonstrates that the juvenile superpredator was a myth and the predictions of future youth violence were baseless. *Amici* have been unable to identify *any* scholarly research published in the last decade that provides support for the notion of the juvenile superpredator, and the scholar credited with originating that term has acknowledged that his characterizations and predictions were wrong; he is one of the *amici* who submit this brief."

The court ruled, narrowly, that it is in fact cruel and unusual to sentence minors to life without parole. This was just one of the twenty-first-century SCOTUS rulings that have scaled back harsh juvenile justice laws. Other rulings eliminated the death penalty for juveniles and mandatory life sentences for crimes that didn't involve killing.[62] These rulings were all part of a broader movement, informed by recent evidence from developmental psychology and neuroscience, to bring the justice system more in line with what we know about young people: They are more impulsive, less able to weigh the costs and benefits of their actions, and capable of shocking acts of wanton recklessness. Just because someone commits a "superpredatory"-seeming act at sixteen doesn't automatically mean they are irredeemable, or that they suffer from any sort of permanent impairment with regard to their morality or humanity.

It was in 2001 that DiIulio acknowledged, in an interview with the *Times*,[63] that he had simply been wrong—though he also claimed, not entirely credibly, that he "couldn't write fast enough to curb the reaction" to his idea, which he described as spinning out of control despite his best efforts to control it. "I'm sorry for any unintended consequences," DiIulio told the *Times*. "But I am not responsible for teenagers' going to prison." This part of the story feels unfinished: Why did DiIulio, a careful academic in other respects, make such spectacular, scary claims, even as he failed to fully define his key term or publish peer-reviewed research on superpredators? He doesn't appear to have given a full account, and it would be interesting to find out; I asked to speak with him through his department at the University of Pennsylvania and was told that due to health problems he wasn't doing interviews.

As for the broader lessons here, two major points stand out. First is the danger that can arise when respected public intellectuals slip into the roles of thought leaders. When an idea is stamped with BELIEVED BY IVY LEAGUE EXPERTS, that is often enough to spur its spread, especially if the idea fits a well-worn cultural or societal groove. The fact that sometimes Ivy League experts are wrong, and sometimes they venture well outside their academic lanes, often gets ignored. Journalists and policy makers often lack the time or the expertise to, for example, trace a given figure (say, that 6 percent statistic from earlier in this chapter) back to the source research and ensure it truly holds water.

The superpredator error also highlights the need for people in positions of power to push back against conceptual fuzziness and insist that terms, particularly terms that are novel or emotionally arresting or both, be defined in a careful, coherent, falsifiable way. When one reviews the history of the superpredator concept, it is remarkable how under-defined it was and how few tough questions its chief architect was asked to answer. When terms are thrown around in a loose manner, when they are taken to mean one thing on Monday and another on Tuesday, that is almost always an indication that half-baked behavioral thinking is occurring. And few instances of half-baked behavioral thinking did more damage than the superpredator meme.

# 3

# OF POSING AND POWER

There's an obvious reason the self-help movement, which grosses about $10 billion per year in the United States, has been such a runaway success: many people feel powerless. Everyday life presents us with an endless array of situations in which we feel batted around by forces beyond our control: economic downturns rob us of income; a spouse unexpectedly walks out on our decade-long marriage; a sudden pandemic turns life upside down; we have to give a presentation early in the morning after our kid was up all night vomiting.

It's unsurprising, then, that there's such a huge market for authors and speakers who tell people, "You have more control than you think." This sentiment has taken many different forms over the years, but in recent decades the self-help scene has shifted from the mushy self-actualization narratives so prevalent in the 1960s and toward the imperative to "buck up"—a trajectory that dovetails neatly with an increasingly cutthroat, competitive labor market. Spencer Johnson's bestselling *Who Moved My Cheese?* from 1998 is a nice exemplar of this subgenre.[1] It advised workers affected by the economic upheavals of the late twentieth century to learn from a pair of maze-dwelling mice who, upon finding out that their reliable daily cheese supply has been moved, begin searching for more cheese without complaint. Another duo, two little humans named Hem and Haw, are also trapped in the maze, but they initially waste their time kvetching about how unfair it is their cheese has been moved. They were not the role models you wanted to follow if you were a laid-off real-life human hoping to survive the tumult of American deindustrialization and downsizing, though they do eventually get wise and realize they must take the initiative and, like their murine counterparts, simply find more cheese.

Body language has often been an important part of the buck-up formula; stop slouching or broadcasting other negative signals, the thinking goes, and people will take you a lot more seriously. There are countless books on the subject with titles like *Body Language: A Guide to Mastering Body Language to Become More Successful, Attractive, and Desired* and *Body Language Training: Attract Women and Command Respect by Mastering Your High Status Body Language*. Whatever you project out into the world, the world will reflect back at you. Or that's the thinking, at least. Suffice it to say that many of these books aren't exactly scientific. Claims about body language and self-help often feel so intuitively correct it's easy to get people to believe them even if you don't present solid supporting evidence.

In 2010, though, the idea that body language might be a secret to success achieved a new level of scientific respectability in the form of a paper published in *Psychological Science*. The co-authors, Dana

Carney and Andy Yap, then of Columbia University, and Amy Cuddy of Harvard, had forty-two students arrange themselves, for two one-minute spans, in either "high power" poses, which involved expansive, assertive positions (leaning back in a chair with their hands behind their head and their feet on a table, or leaning forward over a table with their hands and arms spread out a bit in front of them, their hands touching the table), or "low power" ones, which looked more passive and fearful (sitting in a chair with their hands in their lap or standing in a mummy-like position, their arms wrapped around their bodies and their ankles crossed).[2] Then the students were asked to play a game: They were given $2 and presented with the opportunity to bet it, via a die roll, for a 50 percent chance to win $4 and a 50 percent chance to walk away with nothing. They were also asked how "powerful" and "in charge" they felt on a 1–4 scale. In addition, they provided saliva samples so that the researchers could measure their hormone levels.

The researchers found that those assigned to the high-power poses were more likely to accept the double-or-nothing bet than those in the low-power conditions, and reported feeling more powerful than members of that group. Even more intriguing, the researchers reported the high-power posers had elevated levels of the male hormone testosterone, which is associated with competitiveness and performance, and reduced levels of the stress hormone cortisol, relative to the low-power group. "That a person can, by assuming two simple 1-min poses, embody power and instantly become more powerful has real-world, actionable implications," the authors noted in the abstract. And it would be hard to overstate the impact of both this paper and the many "sequels" it spurred as researchers probed the promise of power posing.

It was Cuddy, by far, who became the most prominent advocate of this idea. In 2012 she gave a TED Talk that would come to be one of the most highly viewed of all time—more than fifty-eight million views, as of August 2020, putting it in second place. In it, she extended the claims from the paper even further. "So two minutes lead to these hormonal changes that configure your brain to basically be either assertive, confident and comfortable, or really stress reactive and feeling sort of shut down," Cuddy, presenting confidently, explains. "And we've all had the feeling, right? So it seems that our nonverbals do govern how we think and feel about ourselves, so it's not just others, but it's also ourselves. Also, our bodies change our minds." That's one of the crucial claims of power posing: improved body language can be beneficial not only by sending different signals to others but by, in effect, sending different signals to one's own self. It's a provocative twist on a pretty old idea.

Cuddy has an inspiring biography. In 1992, when she was in college, she was ejected from a car in an accident that left her with a traumatic head injury. At first the injury profoundly hindered her cognitive abilities. "Recovery from a [traumatic brain injury] is frustrating, confusing, isolating, and never really ends," she wrote in a 2017 Facebook post marking the twenty-fifth anniversary of her accident.[3] In her 2015 bestseller, *Presence: Bringing Your Boldest Self to Your Biggest Challenges*, she writes movingly of her slow, uncertain rehabilitation, of having to relearn, in a sense, how to do complicated, abstract forms of thinking. "So you are probably not going to finish college," one doctor told her shortly after her accident, Cuddy recalled onstage at an event in Brooklyn.[4] He quickly added, "That's okay, because you're going to be high functioning." Cuddy, who at nineteen believed her intelligence was just about the only thing she really had going for her, wasn't exactly reassured. But through a long, arduous recovery process, she was able to earn a PhD in psychology from Princeton, where she worked on issues pertaining to intergroup discrimination and stereotyping under the legendary social psychologist Susan Fiske. After that, she attained a position as a young scholar at Harvard Business School, where she co-authored the power-posing paper.

In large part because of the runaway success of Cuddy's TED Talk, soon power posing was everywhere, with Cuddy its charismatic chief evangelist. It garnered coverage in *The New York Times*, where one Style section profile noted that the idea "promises personal transformation with nary a pill, cleanse or therapy bill."[5] CBS confidently told its viewers, "Believe it or not, her studies show that if you stand like a superhero privately before going into a stressful situation, there will actually be hormonal changes in your body chemistry that cause you to be more confident and in-command," with the segment emphasizing that "Cuddy's work is grounded in science."[6]

Power posing fit neatly into the established self-help niche in American life. Cuddy, of course, was far from the first to promise "personal transformation" premised on a simple, intuitively sensible-sounding idea. But it's usually slick-haired inspirational-speaker types making these promises. The purveyors of these ideas haven't, generally speaking, been Harvard professors.

THAT'S NOT TO SAY Cuddy and her colleagues' work was without academic pedigree. It fit within a constellation of studies in the field of "embodied cognition" that deal with the purported psychological effects of physical touch and movement more broadly. Many of these findings, like power posing itself, gained some mainstream attention because they were so striking and interesting. Take the "Macbeth effect" discovered in 2006. The researchers Chen-Bo Zhong and Katie Liljenquist "asked participants to recall in detail either an ethical or unethical deed from their past and to describe any feelings or emotions they experienced" at the time. The participants who thought about their unethical behavior, the authors wrote, experienced "increased mental accessibility of cleansing-related concepts, a greater desire for cleansing products, and a greater likelihood of taking antiseptic wipes. Furthermore, we showed that physical cleansing alleviates the upsetting consequences of unethical behavior and reduces threats to one's moral self-image." One third of the participants in the "ethical recall" scenario took the wipes they were offered, compared with two-thirds of participants in the "unethical recall" scenario; of those who cleaned themselves with the wipes, just 40.9 percent "volunteer[ed] without pay for another research study to help out a desperate graduate student," while among those who did not clean themselves, almost three-quarters did.[7] The clear implications of this finding were that when the respondents felt morally unclean it led them to feel *physically* unclean, and that the act of cleaning brought with it a sense of discharged moral responsibility.

Older, and more famous, was a group of studies in which participants were asked to hold pencils or pens in their teeth in a manner that forced their faces into a smile. The grinners reported that cartoons were funnier than did those who frowned, and offshoot experiments suggested facial expressions could affect mood in other ways as well. This "facial feedback" finding became so well-known that it was printed in many introductory psychology textbooks and was often presented as scientific evidence supporting a belief that had been around for a long time—that our bodies shape our feelings, rather than the other way around. That is, maybe we get happy because we smile, rather than smile after getting happy. Cuddy mentioned these studies in her TED Talk as an example of how our bodies affect our minds.

Some embodied cognition studies have been hit particularly hard by the replication crisis. In one 2014 study, a team of researchers was unable to replicate the Macbeth effect, despite attempting multiple times, with more statistical strength than the original study.[8] Worse, as Daniel Engber explained in a comprehensive 2016 article in *Slate*, a carefully constructed attempt to replicate the pencil finding coordinated by the psychology researcher E. J. Wagenmakers—an effort carried out in "17 labs spanning eight countries"—didn't go well. "When Wagenmakers put all the findings in a giant pile, the effect averaged out and disappeared," wrote Engber. "The difference between the smilers and frowners had been reduced to three-hundredths of a rating point, a random blip, a distant echo in the noise."[9]

Power posing's own replication travails started in 2015. By then, a handful of follow-up studies had been published, not all of them telling as straightforward a story as the original. In one 2013 study published in the *European Journal of Social Psychology*, for example, those in high-power poses who, via a laboratory trick, were made to feel excluded, felt the sting of that ostracization *more* keenly than those in low-power poses (who felt about the same whether they were included or excluded).[10] Other studies, though, did seem to point in the same general direction as the original blockbuster finding. A 2012 paper in the *Journal of Experimental Social Psychology*, memorably titled "It Hurts When I Do This (or You Do That): Posture and Pain Tolerance," found that power posers had a higher pain threshold than those who adopted low-power poses.[11]

But in March 2015, a team led by Eva Ranehill published the results of an attempted replication of the first power-posing study that employed a sample size of two hundred—about five times larger than the original sample—and that "closely followed" its methodology.[12] Ranehill's team found no effects whatsoever when it came to the correlation between power posing and either risk taking or hormone levels, directly contradicting the original finding and some of the promising-looking follow-up studies that had been published in the interim. That failed replication, in turn, greatly amplified a mainstream media conversation about the replication crisis that was, at that point, still in its infancy; in *Slate*, for example, Andrew Gelman and Kaiser Fung wrote that while it was understandable that "outsiders" were fooled by all the glowing coverage of what appeared to be a thin finding, "insiders who are aware of the replication crisis in psychology research are suspicious of these sorts of dramatic claims based on small experiments. And you should be too."[13] (We'll take a deeper look at the replication crisis in chapter 7.)

It was one of Cuddy's co-authors on the original 2010 study, though, who truly kicked the legs out from under power posing. In September 2016, Dana Carney posted a document to her UC–Berkeley faculty webpage in which she wrote, underlined and in bold, "I do not believe that 'power pose' effects are real."[14] Carney then elaborated in rich, fairly damning detail: she believed that she and her colleagues had engaged in a statistical technique that, while commonly employed at the time of the original power-posing study in experimental psychology (and other fields), is now known to greatly improve the probability of detecting phenomena that aren't really there.

It's called *p*-hacking, and it's a central concern of the reformers presently worried about the state of psychological science. To understand *p*-hacking, it's important to understand how scientists have traditionally defined a "significant" result: by reference to its so-called *p*-value. This involves what is basically a two-step process. First, you say, "Let's suppose that there's nothing here"—that is, power posing doesn't do anything, or the pill doesn't work, or this educational intervention has no effect. Scientists refer to this as accepting the "null hypothesis." Then you ask, "If the null hypothesis is true, what is the probability of nonetheless finding a difference in the data at least as large as what is being observed here?" So for a study involving one group that power poses, a control group that doesn't adopt any pose, and measures of testosterone as the outcome variable of interest, you'd say, "If power posing doesn't do anything, what is the probability we'd nonetheless observe a difference at least this large in the average testosterone levels between the power-posing group and a control group?"

Because of the nature of randomness, sometimes mere statistical noise can look like a pattern. If a non-posing group has less testosterone than the power-posing group, that could be because of the effects of power posing, or it could be a random blip. A simple hypothetical can drive this home more forcefully: Consider the example of flipping a coin repeatedly to attempt to gauge whether it is fair (while setting aside the compelling and interesting argument that it actually isn't possible to weight a fairly flipped coin in this manner[15]). Over many flips, a fair coin is going to approach a ratio of one heads for every tails. Along the way, one may well encounter suspicious-seeming sequences in which the ratio is askew, or when strings like HHHHHH or even TTTTTTTTTTTTT are observed. But such sequences are not *alone* evidence of an unfair coin; randomness can deliver the occasional streak that looks meaningful but isn't. The point of a *p*-value is to account for this; to publish a result with a low *p*-value is to say "not only did I observe this difference, but for statistical reasons it is very unlikely to be the result of random noise." In the context of a coin, one would need to encounter a very skewed ratio of heads to tails, or vice versa, over *many* tosses to generate a *p*-value that indicates the coin is probably unfair.

So imagine you run a study of power posing and observe a statistically significant testosterone difference between power posers and a control group; the power posers have more T. If your number crunching spits out a value of $p = 0.04$, that means, in theory, that there is only a 4 percent chance you would have observed a difference at least this large between the groups if power posing didn't do anything. This doesn't mean power posing *caused* the difference—recall that crucial distinction between causation and correlation—but if you structure the experiment the right way, you can effectively rule out most other potential causes for any such differences that are observed. By convention, results in psychology and many other sciences are viewed as statistically significant, and therefore publishable, if their *p*-value is 0.05 or lower. In other words, they are publishable if there's a 5 percent or lower chance a result at least that large would have occurred if the null hypothesis were true. This really is an arbitrary number, and as we'll see later, some researchers have argued it should be bumped downward to make science more rigorous (or an entirely different type of statistical test should be used), but those are the current rules that prevail in most of academic publishing.

*P*-values are very manipulable; that's where *p*-hacking comes in. By including or excluding trials (whether coin flips or samples of salivary testosterone levels), one can juke a stubbornly too-high *p*-value down below that critical $p = 0.05$ threshold, at which point it is—*ta-da*—publishable. This is what Carney admitted she, Cuddy, and Yap had done in their original finding. By many accounts, it was a common practice at the time; the risks of generating false-positive findings weren't well understood by most research psychologists. So the image here should be not of evil researchers conniving to publish bad research but of excited and perhaps somewhat quantitatively under-informed researchers falling into a statistical trap. And later Carney realized what had happened. "The data are flimsy," she wrote on her site. "The effects are small and barely there in many cases." Hence, "I do not have any faith in the embodied effects of 'power poses.' I do not think the effect is real." She also pointed out, "I discourage others from studying power poses." That document has been all that she has said on the matter publicly, however. When I asked Carney if she'd talk about it, she politely declined.

Soon after Carney publicly turned against power posing, two other researchers presented another major challenge to it. In 2017, the top journal *Psychological Science* published an article by Joseph P. Simmons and Uri Simonsohn that employed an ingenious statistical tool they had developed to evaluate bodies of scientific literature from a bird's-eye altitude.[16] Known as a *p*-curve, "it tells us whether a set of statistically significant studies is actually true/replicable, or whether those significant results were caused by *p*-hacking or file-drawering," Simmons explained in an email. (If twenty studies are conducted on a given question, eight find strong evidence and are published, three find moderate evidence and are published, and nine find no evidence and aren't published—they are thrown in a "file-drawer" and forgotten—then a survey of the *published* studies will offer a distorted picture of the evidentiary landscape. This is why the best and most thorough meta-analyses seek out both published and unpublished studies.)

Glossing over some of the gory statistical details: If there's a real effect being observed—in power posing or anything else—a graph of the *p*-values for this body of research should have a specific shape to it: many more *p*-values between 0 and 0.025 than between 0.025 and 0.05. If, instead, many studies are clustered right below $p = 0.05$, that's a strong sign the literature is weaker than it appears due to *p*-hacking, file-drawering, or both.

And when it came to the studies Carney, Cuddy, and Yap compiled to show that power posing had a strong base of evidence behind it (before Carney's post, of course), Simmons and Simonsohn indeed found a suspicious number of studies clustered right below that $p < 0.05$ threshold. They wrote that this *p*-curve, combined with the failed replication Ranehill's team had published, "suggest[s] that the behavioral and physiological effects of expansive versus contractive postures ought to be treated as hypotheses currently lacking in empirical support."

In response, Cuddy and power posing's other remaining defenders pivoted a little and began focusing more on the claim that power posing can increase one's self-*perception* of power than on the hormone and behavior claims. It's possible this is true: one 2017 meta-analysis turned up "moderate" evidence for the felt-power effects of power posing on subjects unfamiliar with the idea that power posing increases felt power (and who therefore wouldn't be susceptible to placebo effects).[17] Of course, if power posing has this effect only on people who aren't familiar with the concept, then the act of informing them takes away its power altogether, which is a clear problem. Either way, if you enjoy power posing or think it helps improve your performance or whatever else, there's no reason not to do it: it's unlikely to be harmful, as long as you don't overestimate its likely influence.

But power posing became a worldwide phenomenon because of various very impressive-sounding claims, including claims that it can rewire your brain to jolt you from passivity into a sense of power—not because it simply offers a mild potential boost to your mood. Overall, while her focus has shifted, Cuddy hasn't fully backed off her view that there is compelling evidence power posing works in the way she originally described. In April 2018, she, Jack Schultz of Harvard Business School, and Nathan Fosse of Harvard's Institute for Quantitative Social Science, released their own *p*-curve analysis in *Psychological Science* that, they argued, showed "strong evidential value for postural-feedback (i.e., *power-posing*) effects and particularly robust evidential value for effects on emotional and affective states."[18] But as Simmons, Simonsohn, and Marcus Crede, a psychologist we'll meet more in depth later, have noted, the papers they included had some rather crippling problems, ranging from *p*-values so low they are inherently suspicious (in one case, $p = 0.00000000005$, or "1 in 2 trillion")[19] to results that didn't actually show that power posing had positive effects relative to neutral poses, but rather that slouching had *negative* effects, which is a different claim. So these results still don't really support Cuddy's bold, early claims about power posing. (Crede would later publish another paper arguing that effectively *all* positive power-posing results in the literature came from comparisons of neutral poses to slouch-y ones, not the more relevant comparison of neutral poses to power poses.[20])

Power posing's failed replications have received public notice—in *Slate*, in Anand Giridharadas's bestselling book, *Winners Take All*, and in a *New York Times Magazine* profile that was fairly sympathetic to Cuddy. And yet it doesn't seem the research that calls power posing into question has gotten nearly as much public attention as the early claims in its favor. That, as always, is the problem.

WHEN ONE CRITIQUES HALF-BAKED behavioral science claims, it's important not to leap too far in the other direction. Half-baked science is often based on at least some truth; the problem is less the kernel than what's made of it. As we've seen so far, it's true that one's upbringing affects one's propensity for criminality; it's false that massive numbers of young people are turned, by their circumstances, into immoral monsters incapable of distinguishing right and wrong. It's true that if you view yourself as completely worthless and incapable of being loved or accomplishing anything, this mindset could hinder you; it's false that there is a straightforward causal relationship between self-esteem and most important life outcomes.

When it comes to power posing, it's true that body language can have meaningful effects on human social life and that knowledge of the specific ways it does so in a given culture could be beneficial. Of course, Cuddy's claim is different—she says that your body language affects how you feel about *yourself* so significantly it could have profound personal implications ("two minutes lead to these hormonal changes that configure your brain to basically be either assertive, confident and comfortable, or really stress-reactive, and feeling sort of shut down")—but it's still important not to disregard the general principle.

It goes without saying that in the animal kingdom body language matters a great deal. Dogs, as everyone knows, indicate submissiveness or playfulness by rolling over on their backs. Cats arch their backs in a very distinctive way when they are unhappy with or threatened by another creature. In the wild, massive animals' epic-seeming fights over mates or territory can end in an instant when one of the participants flashes a subtle cue indicating surrender. Body language is everywhere.

As a general rule, when a behavior is everywhere among animals, it stands to reason that humans have it too, and of course humans indicate certain things with body language. When I first began researching this chapter, I watched a then-recent NBA playoff game between my Boston Celtics and the Cleveland Cavaliers. In one sequence, the Celtics player Jaylen Brown cut to the hoop, received a nice pass from Al Horford, was fouled by the Cavaliers' J. R. Smith (a player with something of a reputation for hotheadedness), and successfully converted the somewhat wild runner he tossed up as the foul occurred.[21] I remember watching the game with my brother and thinking the foul looked dirty—that Smith threw a forearm that hit Brown in the side of the head.

But watching Smith's body language immediately after he committed the foul, in slow motion, made me change my mind; at the very least, I no longer thought it was an *intentionally* dirty play. In the clip, Smith does two things immediately after making contact with Brown: he holds a palm out to Brown and he takes a few slow steps toward him. It isn't clear what, if anything, he says to Brown, but his body does the talking for him: some combination of *My bad* and *Sorry about that* and *That wasn't intentional*. It's strange, the potency of those two seconds of body language: it looks like an immediate, instinctive act designed to broadcast remorse and avoid conflict, and it has the effect of instantly changing the observer's interpretation of the physical contact that just occurred—and, it stands to reason, it may have had a similar effect on Jaylen Brown.

So yes, body language can communicate important information between humans. This is common sense. It's also been studied and theorized about for a long time by multiple disciplines, perhaps most famously by the legendary sociologist Erving Goffman, whose "dramaturgical model" argued that everyday social life should be understood as an endlessly unfolding series of performances. This is true even for seemingly straightforward interactions, like going to a pizzeria. I might present myself as a polite, easy customer in the hopes of getting better service. The cashier, if his boss is present and he is new, might present himself as particularly enthusiastic and competent in the hope of picking up extra shifts: he'll move around crisply and efficiently and place my slice gently down on the counter next to me. If his boss isn't there, however, he might present himself differently, acting in a surly manner to communicate to me that while I am the customer, that doesn't mean I'm better than him or deserve any sort of special treatment: he'll move sluggishly to and from the oven, toss the slice at me from a few inches above the counter rather than gently set it down, and so on. And body language is an important part of these performances; from body language alone, I can make some likely accurate inferences about how the guy at the slice joint is going to treat me as a customer.

Taking stock of our performances, and how they sometimes go awry, Goffman noted that "unmeant gestures" can acquire a "collective symbolic status." For example,

> a performer may accidentally convey incapacity, impropriety, or disrespect by momentarily losing muscular control of himself. He may trip, stumble, fall; he may belch, yawn, make a slip of the tongue, scratch himself, or be flatulent; he may accidentally impinge upon the body of another participant. Secondly, the performer may act in such a way as to give the impression that he is too much or too little concerned with the interaction. He may stutter, forget his lines, appear nervous, or guilty, or self-conscious; he may give way to inappropriate outbursts of laughter, anger, or other kinds of affect which momentarily incapacitate him as an interactant; he may show too much serious involvement and interest, or too little.[22]

A lot of this is intuitive. Within a given culture, just about everyone learns, from a young age, which sort of body language communicates respect and deference, how to tell when someone is getting bored by your story about watching a YouTube video of a random NBA play, and so on. Body language is clearly a big part of our everyday lives. The unpleasantness of the "close talker" character in *Seinfeld*, a boyfriend of Elaine's who stands far too close to people when he speaks to them, resonates with just about everyone, because just about everyone would feel uncomfortable if a conversation with a new acquaintance entailed flecks of spittle and an olfactory tour of their last meal.

But do such instinctive perceptions matter? Do they shape our behavior over time? In fact, there's some evidence that once these first impressions of others are formed, they have a strong so-called anchoring effect on our future evaluations. When I interviewed the Columbia Business School psychologist Heidi Grant Halvorson about first impressions in 2015, she told me, "I always want to be able to say to people, 'There's this really easy thing you can do.' But the truth is, the more you know about person-perception, the more you realize how hard it is to change first impressions."[23] And first impressions often aren't based on much substance. As Malcolm Gladwell put it in *Blink*, "Snap judgments are, first of all, enormously quick: they rely on the thinnest slices of experience."[24] Though it would be hard to express the practical ramifications of this more colorfully than Dwight Schrute did in *The Office*: "I never smile if I can help it. Showing one's teeth is a submission signal in primates. When someone smiles at me, all I see is a chimpanzee begging for its life."

In the real world, if I meet you and quickly determine you are a jerk, it will be pretty difficult—though by no means impossible—for me to be swayed from that view. I'll likely view our subsequent interactions through the lens of "s/he seems like a jerk"; confirmation bias will kick in, a bit. Body language, both research and common sense tell us, is part of this: if you immediately get too close to me, looming over me like a drill instructor, and shake my hand with far more force than is necessary, of course that will contribute to the impression I am forming of you. This is all useful information for people to know.

The problem is the leap from modest claims that are justifiable on the basis of science (or common sense) to big claims that aren't. Cuddy found, among other things, differences in hormone levels between high- and low-power posers, and differences in their appetite for participating in a small betting game. From this razor-thin evidence, she extrapolated to much stronger, less empirically defensible assertions.

MAYBE CUDDY'S CLAIMS WERE so striking they would have received widespread attention no matter what. But it's likely that power posing owes some of its success to its fit with certain contemporary feminist currents.

In 2013, about a year after Cuddy's viral TED Talk, Facebook's chief operating officer Sheryl Sandberg's *Lean In: Women, Work, and the Will to Lead* hit American culture like a meteor. The book, which cited Carney, Yap, and Cuddy's study, sparked a clamorous discussion about women in the workplace and America's lingering forms of gender inequity. Sandberg's main argument is that women are all too often socialized into helplessly accepting second-class status in professional settings. To get ahead, they must, well, lean in—assert their rightful place alongside the men who dominate America's boardrooms and C-suites. It was an empowering message, and it resonated to the tune of millions of sales and a sequel of sorts called *Lean In for Graduates*.

Cuddy's *Presence* and Sandberg's *Lean In* both stress individual women's agency; they emphasize, in classic contemporary self-help fashion, that *there are things you can do to improve your station*. Together, they helped launch a particular brand of twenty-first-century feminism that combined many of that movement's traditional concerns with a self-help component that focused much more on individuals than on institutions and social structures. In this model, the problems are societal and organizational (the two often overlap), but the proposed solutions are almost always individualistic.

It's quite striking how similar *Presence* and *Lean In* are on this front; they can both be seen as feminism for an age of fracture, geared at a professional audience. Both almost entirely ignore the possibility of repairing *institutions* in a manner that would make them fairer; instead, it's the job of individual women to act in a more stereotypically masculine way.

That isn't to say that Cuddy and Sandberg don't acknowledge various structural and institutional barriers to women getting ahead; they do, but they subsequently ignore most potential solutions that involve addressing those barriers. Cuddy, for example, notes in her book that misogynistic prejudice is a clear obstacle to gender equity that, "sadly, is not going away overnight. That's not an excuse to ignore the problem, but we're not going to eliminate it tomorrow. I teach psychological research on sexism and racism, but I have very little hopeful news to share about remedies." So the focus isn't on fixing these problems but on helping individuals do what they can to get ahead despite them. "I still actively study the origins and effects of prejudice, but now more than half my research focuses on identifying scientifically grounded mini interventions—things people can do to perform well even when faced with negative judgments and biases," she writes. "Even when the negative judgments and biases are their own."[25] When she discusses collective action, it's collective action with a very particular focus. "I am issuing a challenge to all of us," she writes at one point, "and it's one that I do not take lightly: *Let's change it.* When you see your daughters, sisters, and female friends begin to collapse in on themselves, intervene. Show them examples of girls and women in triumphant postures, moving with a sense of power, speaking with authentic pride."[26] For the most part, "change" is connected not with advocating for different policies or a different distribution of power but with getting individuals to realize the power and potential they are already holding within themselves.

Cuddy's framing seems to imply that there's little to be done about institutions affected by bias. In reality, though, the options aren't "end misogyny and negative stereotypes about women," on the one hand, and "help individual women overcome misogyny and negative stereotypes," on the other. There are many *institutional* fixes that could, in the meantime, render organizations fairer toward women.

Sandberg's book contains many missed opportunities on this front. At one point, for example, she mentions that at Harvard Business School class participation accounts for half of students' grades, and "professors teach ninety-minute classes and are not allowed to write anything down, so they have to rely on their memory of class discussion."[27] (Or at least this was the case when she was a student there.)

A situation like this clearly incentivizes a certain form of bullshit artistry that, it could be argued, favors men and male socialization. Professors are probably more likely to remember confident, skillfully delivered proclamations than more nuanced, hedged remarks, and woe be unto students who are naturally introverted and who don't always have something to say in the heat of the moment, perhaps because they have a penchant for fastidiousness and prefer to triple-check an answer before publicly proclaiming it. In the world of leaning in and power posing, the solution to a conundrum like this is to teach women to perform this sort of traditionally male confidence and self-promotion—both Sandberg and Cuddy specifically extol "fake it until you make it" principles—rather than to question why these rules are in place at all or to advocate for reforming them. But why should Harvard Business School professors never write anything down and instead rely on their memories when determining students' grades, given how fallible memory is and given the perverse incentives this likely introduces? "End misogynistic discrimination" might be an out-of-reach goal at the moment, but is "change nonsensical grading practices that unduly reward bluster"?

Later in the book, Sandberg writes, "An internal report at Hewlett-Packard revealed that women only apply for open jobs if they think they meet 100 percent of the criteria listed," compared with apparently overconfident men, for whom the cutoff point was about 60 percent.[28] Sure, one response to this finding would be to encourage women to ape men and apply to jobs they aren't, strictly speaking, qualified for. But why not instead address HP's strange apparent practice of inviting under-qualified men in for interviews in the first place? It doesn't make sense to list "criteria" for a position and to then simply ignore the fact that some applicants don't meet those criteria. Why not simply follow the rule to the letter, or rewrite the job descriptions so as to make them more accurate and less likely to generate a skew in which overconfident people are more likely to apply? If you enforce the requirements as written, it *won't matter* if male employees are more confident than female ones in your organization, because only those who fit the requirements will advance to the interview stage anyway.

Power posing preaches that social and institutional structures might seriously be flawed, but we're more or less stuck with them, so we should spend most of our effort helping women succeed in the sorts of situations where, to take the two above examples, superficial bloviation and overconfidence are rewarded. In a sense, this idea is subtly baked into the experiment that first launched power posing. Recall that students were given $2 and asked whether they wanted to risk it for a coin's-flip chance to go home with $4. The authors describe this as "a risky but rational bet," but it's unclear why. Typically, economists view one choice as more rational than another if it is associated with a higher expected value, which is calculated by multiplying the expected gain by the probability it will occur. In this case the two choices have the same expected value: $2. That is, if you make one choice, you have a 100 percent chance of keeping your $2, or $1.0 \times \$2 = \$2$, and if you make the other, you have a 50 percent chance of getting $4 and a 50 percent chance of getting nothing, or $(0.5 \times \$4) + (0.5 \times 0) = \$2$. So while this isn't exactly a high-stakes situation, taking the bet is *not*, by the normal definition, a rational choice: it would be more accurate to say the two decisions are equally rational.

Cuddy and her colleagues claimed it's a good thing that their intervention nudged the participants to take this risk, and this claim went mostly unexamined. But is it a good thing? Is that the sort of "feeling powerful" we want to promote? Does American society, and the American economy—both still dominated by males, at least at the highest levels—suffer from a dearth of pointless risk taking?

IT WOULD BE NICE IF, in lieu of power posing and other, similarly individualistic offerings, there were some menu of highly proven, effective interventions on offer to counter gender bias. Alas, this is a complicated area with few easy answers and a particularly high potential for unintended consequences; for instance, some research suggests that women are *punished* for acting more confidently, even as men are rewarded for it, which could mean that even if power posing worked, it wouldn't necessarily lead to good outcomes. (This and other possibilities of "unintended consequences of gender bias interventions" are mentioned in a very thoughtful paper by Suzette Caleo and Madeline E. Heilman published in *Archives of Scientific Psychology* in 2019.[29])

There are also questions to raise about Cuddy's basic assumptions. While there are countless examples of women feeling powerless, talked over, and disenfranchised in work settings, Cuddy's diagnosis that gender disparities stem at least in part from women not feeling powerful enough in the workplace is by no means bulletproof: as the researcher Laura Guillen argued in *Harvard Business Review* in 2018, the idea of a male-female "confidence gap" in the workplace isn't really supported by the literature.[30] (While power and confidence aren't quite the same thing, the latter certainly feels like a solid enough proxy for the former.) If even basic questions like this remain in dispute, it's no wonder that researchers haven't yet come up with reliable ways to effectively address these issues.

Making matters worse, research on gender bias has by no means been immune to the methodological problems roiling behavioral science, including both failed replications and a growing number of findings that, upon a closer examination of their original data, don't quite stand up to scrutiny. Even the most solid-seeming results can turn out to be quite shaky. One of the most influential, textbook findings in this area, for example, is a 2000 study by the economists Claudia Goldin and Cecilia Rouse of bias in orchestra auditions. They claimed that instituting a screen in orchestra auditions—so that the juries who determine new hires could hear but no longer see the auditioners—greatly ameliorated what had been significant bias favoring male candidates. "We find that the screen increases—by 50 percent—the probability that a woman will be advanced from certain preliminary rounds and increases by severalfold the likelihood that a woman will be selected in the final round," they wrote.[31]

In 2019, the independent data scientist Jonatan Pallesen published a blog post casting serious doubt on this result.[32] That post found its way to Andrew Gelman, the Columbia political scientist and blogger mentioned earlier in this chapter who is known as one of the best quantitative detectives around. Gelman examined the study, found no evidence anywhere for the 50 percent claim, and, more broadly, described the results as "not very impressive at all"—riddled with statistical noise.[33] He even cited another researcher's analysis that when everything was controlled for, women actually did *worse* in the blind auditions.

If we can't trust as respected a finding as this one, it should cause us to be more skeptical of these sorts of interventions in general. Which leaves us where, exactly? Gelman himself provides an answer. Even after expressing a great deal of doubt about the Goldin/Rouse study, he writes, "I agree that blind auditions can make sense—even if they do not have the large effects claimed in that 2000 paper, or indeed even if they have no aggregate relative effects on men and women at all."

Maybe that's the key lesson here: there are certain actions institutions can take that, at the very least, are unlikely to cause harm and may be capable, at least theoretically, of attenuating the impact of certain sorts of biases. If you start encouraging professors to take notes to avoid the need to remember who said what, that is clearly going to make their recollections more accurate, reducing the chance that faulty recall will introduce bias into their grading. If you institute a system in which requirements for promotions are strictly adhered to, ditto. Or take interviews themselves: There is solid research in the field of industrial-organizational society that suggests that traditional, so-called unstructured interviews are of approximately zero use to predicting whether a new employee will perform well.[34] And that's exactly the sort of situation where, especially in institutions that are already male dominated, bias could rear its head one way or another. It's only human to feel a warm glow toward someone who comes from a similar background as you (the impulse illustrates the general phenomenon of homophily, or our tendency to like those who are similar to us). In the case of professional organizations dominated, at the management level, by men, that's a recipe for biased hiring. So why not get rid of unstructured interviews?

Other institutional changes might be harder to implement but could also address gender gaps effectively. Goldin, for example, has pointed out that the easier it is to replace one employee with another in a given task or customer interaction, the easier it is to offer the sorts of flexible hours that help women continue to work, and advance, even if childcare responsibilities fall disproportionately on them and force them to reduce their hours. (She has identified those responsibilities as a key cause of certain gender disparities in the workplace, given that most women have children during crucial early-career years.) The field of pharmacology illustrates the point. While it requires a great deal of training and expertise, and pays pretty well, consumers don't really care if "their" pharmacist or another one is filling out their prescription.[35] As a result, women can continue to advance even if they sometimes reduce their hours. Goldin believes that if other fields, like the law, could increase their substitutability along these lines, this could help reduce their wage gaps—and pharmacy in fact performs quite well on this front.

While the empirical literature on professional gender disparities is rather fuzzy and lacking in straightforward answers—here Cuddy is correct, even if she is overconfident about her own preferred intervention—it does offer some hints about how progress could be made. And those hints point toward changing institutions, not individuals.

# 4

# POSITIVE PSYCHOLOGY GOES TO WAR

There's a story Daniel Kahneman, the Nobel Prize–winning psychologist who is responsible for many of the most trenchant modern insights into human folly, likes to tell in interviews. It's also in his book *Thinking, Fast and Slow*:

> Many years ago I visited the chief investment officer of a large financial firm, who told me that he had just invested some tens of millions of dollars in the stock of Ford Motor Company. When I asked how he had made that decision, he replied that he had recently attended an automobile show and had been impressed. "Boy, do they know how to make

> a car!" was his explanation. He made it very clear that he trusted his gut feeling and was satisfied with himself and with his decision. I found it remarkable that he had apparently not considered the one question that an economist would call relevant: Is Ford stock currently underpriced? Instead, he had listened to his intuition; he liked the cars, he liked the company, and he liked the idea of owning its stock. From what we know about the accuracy of stock picking, it is reasonable to believe that he did not know what he was doing.[1]

Part of what makes this a wonderful anecdote is that childlike exclamation: "Boy, do they know how to make a car!" As Kahneman notes, that couldn't be less relevant to the question of whether it made sense to invest a massive sum of money in Ford at that particular moment.

Kahneman's point is that even powerful, otherwise intelligent people can make decisions that feel intuitively correct but that are not based on an accurate understanding of the situation. Call it unskilled intuition; the executive lacked skill when it came to stock picking but went ahead and plowed a fortune into Ford on the basis of a gut feeling that any genuinely competent investor would have ignored.[2] Unskilled intuition affects us all; because as humans our brains often fool us and distract us, we're all susceptible to the Dunning-Kruger effect—the tendency to overrate our own ability in domains where we lack skill.

This occurs in plenty of lower-stakes situations too: when I develop a strong feeling about whether a car mechanic is trustworthy even though I have only the faintest idea of how cars work and can barely understand what he's telling me about my engine light, that's unskilled intuition. But unskilled intuition has the most potential to cause harm—or at least to redirect resources toward dead-end projects—at the level of institutions. There, people in positions of authority are often forced to make consequential decisions in unfamiliar areas on the basis of very little solid information, often not even knowing how to evaluate the available evidence rigorously.

Matthew Grawitch, an organizational psychologist and the director of strategic research in the School for Professional Studies at Saint Louis University, explained this concept nicely in an email to me. "Leaders usually end up in their positions because they are an expert in some aspect of their business," he wrote. "However, when it comes to solving some problem that is less about the product or service being sold and more about more tangential factors (e.g., how to reduce absenteeism, how to reduce turnover, how to make the workplace more psychologically healthy), they are dependent on others—often consulting firms that peddle the latest and greatest panacea. All that has to happen is the firm provides a plausible explanation for how its intervention/approach will fix the problem, and the leader's biases/heuristics and lack of expertise will do the rest."

We've already seen that psychological findings can be true-ish without actually being as relevant to real-world problem solving as they might appear to be at first blush. Applying psychological science often involves translating findings that originate in very specific research situations to the much bigger, messier setting that is the real world. As the University of Virginia legal scholar Gregory Mitchell argues, "What much of the public probably fails to appreciate . . . is that much of the psychological wisdom on offer is based on very limited bodies of research and that much of psychological science places little value on assessing the external validity of research," "external validity" meaning simply the question of whether a research finding is applicable to the real world.[3]

It takes a particularly high degree of skill and knowledge to distinguish a half-baked psychological claim from a genuinely promising one, and it's unrealistic to expect the average human resources manager or school principal or other institutional decision-maker to possess such skill and knowledge. In light of all this, it stands to reason that when these sorts of professionals are handed a menu of psychological interventions available for purchase, unskilled intuition is particularly likely to rear its head.

Applied psychology brings with it certain dangers, then: that claims will be oversold, that money will be wasted on the wrong things, that important people trying to solve pressing problems will be waylaid by zombie theories. Few areas of behavioral science better exemplify these dangers than the increasingly popular endeavor of positive psychology, one of the newest established subfields of psychology and one that has successfully sold questionable theories to many institutions, most notably the U.S. Army, for very large sums.

POSITIVE PSYCHOLOGY WAS FOUNDED by Martin Seligman, a legendary researcher at the University of Pennsylvania (Mihaly Csikszentmihalyi also played a key role and co-authored a pathbreaking 2000 article with Seligman[4]). Marty, as he is known to everyone in his orbit, is an iconic, divisive figure in the world of academic psychology. If you looked only at the early part of his career, you'd be surprised that he ended up being one of the godfathers of a field dedicated to positivity. That early work was, well, dark. Through experiments that involved shocking dogs in different ways, Seligman developed the extremely important psychological concept of learned helplessness. It refers to a situation in which an organism's lack of control over its surroundings prompts it to stop engaging in standard acts of self-preservation. Seligman and his colleagues broadened the idea to humans: certain people, in certain circumstances, more or less *give up*. Soon "learned helplessness" was a textbook concept in psychology.

But later in his career, Seligman shifted focus dramatically. He came to realize, as he would explain in many speeches and interviews, that psychology was too focused on pathology, on fixing broken people, and less on cultivating strengths and helping people who were otherwise basically healthy to maximize their potential. Seligman wanted to transform American psychology, and the best way to do that would be as president of its flagship organization, the American Psychological Association. So in the mid-1990s, he campaigned for that position by arguing, as Rob Hirtz put it in a *Pennsylvania Gazette* article, "for the field of psychology to expand its myopic focus on treating mental illness to include promoting mental health."[5]

Seligman's ideas for promoting mental health have generally centered on individual self-improvement. In a 1975 book, he wrote, "Poverty is not only a financial problem but, more significantly, a problem of individual mastery, dignity, and self-esteem."[6] He has never had much sympathy for common left-of-center beliefs about people being helpless, to a certain extent, before the power of huge structural forces; he views that sentiment as self-pitying. "In general when things go wrong we now have a culture which supports the belief that this was done to you by some larger force, as opposed to, you brought it on yourself by your character or your decisions," he said in a 1999 interview. "The problem about that is it's a recipe for passivity and giving up and helplessness. So one of the things that is going on, I think, is a pervasive victimology." [7]

Seligman's APA campaign was successful. He was elected president in 1996—"by the largest vote in modern history," his faculty page notes[8]—and when he took office in 1998, he made positive psychology the theme of his presidency. This new, highly marketable subfield arrived at an important juncture for the broader discipline of psychology. As Barbara Ehrenreich explained in her 2009 critique of the positive-thinking movement, *Bright-Sided: How the Relentless Promotion of Positive Thinking Has Undermined America*:

> Positive psychology provided a solution to the mundane problems of the psychology profession. Effective antidepressants

> had become available at the end of the 1980s, and these could be prescribed by a primary care physician after a ten-minute diagnostic interview, so what was left for a psychologist to do? In the 1990s, managed care providers and insurance companies turned against traditional psychotherapy, effectively defunding those practitioners who offered lengthy courses of talk therapy. The Michigan Psychological Association declared psychology "a profession at risk" and a California psychologist told the *San Francisco Chronicle* that "because of managed care, many clinical psychologists aren't being allowed to treat clients as they believe they should. They still want to work in the field of helping people, so they're moving out of therapy into coaching." If there was no support for treating the sick, there were endless possibilities in coaching ordinary well people in the direction of greater happiness, optimism, and personal success.[9]

In this view, positive psychology enabled professional psychology to greatly expand its market, allowing psychologists to advertise themselves as coaches to companies, schools, and other organizations. Because positive psychologists didn't diagnose or treat clients in the traditional senses of those terms, there was no need to involve insurance companies; just about everyone was a potential "client."

The global headquarters of positive psychology is Seligman's Positive Psychology Center at the University of Pennsylvania, which was founded in 2003. Separate from the school's traditional psychology department, the Positive Psychology Center offers a one-year Master of Applied Positive Psychology degree; the program trains students in the skills they need to work professionally as positive psychology coaches and consultants in what has become a burgeoning field. (MAPP graduates can't practice clinical psychology or counseling as such.)

At root, positive psychology claims that there are reliable ways to make people happier and more optimistic and that these changes bring with them benefits like increased mental health and longevity. But the underlying science is controversial. As Daniel Horowitz writes in his excellent 2018 cultural and scientific history of the field, *Happier? The History of a Cultural Movement That Aspired to Transform America*, "Virtually every finding of positive psychology under consideration remains contested, by both insiders and outsiders . . . Major conclusions have been challenged, modified, or even abandoned."[10]

Certain crucial issues regarding the potential for correlational confusion—do people live longer because they're happy, or are they happy because they are healthy and therefore live longer?—haven't been fully addressed, and these and other questions are sometimes ignored in favor of pat story lines. Some positive psychologists, for example, long claimed a link between an individual's level of happiness and the strength of their immune system that had only weak evidence behind it. That claim isn't heard as much these days but it was echoed widely, including by Seligman himself.[11]

One of the most outlandish claims to come out of positive psychology is the idea that there exists a "critical positivity ratio" of positive to negative thoughts strongly correlated with well-being. As two critics describe this idea, which was first posited by Barbara Fredrickson and Marcial Losada, the ratio "was supposedly derived from the famous Lorenz equation in physics by using the mathematics of nonlinear dynamic systems, and was defined precisely as '2.9013.'"[12] Fredrickson and Losada went as far as to argue that this precise figure is invariant across different cultures, races, and genders—a true human universal. Seligman fully embraced and helped to promote some of these claims. "The first time I heard Barb Fredrickson speak, the famous psychologist sitting next to me said, 'That's the real thing!'" he said in a glowing endorsement of her book *Positivity: Top-Notch Research Reveals the 3 to 1 Ratio That Will Change Your Life*. The book, he promised readers, is "the perfect blend of sound science and wise advice on how to become happier. Barbara Fredrickson is the genius of the positive psychology movement."[13]

Another, superficially more plausible, formula offered by positive psychology is also rather shaky, and was also boosted enthusiastically by Seligman. In a highly cited 2005 article, the positive psychologist Sonja Lyubomirsky and her colleagues argued that 50 percent of the variance in human happiness is accounted for by genetics, 10 percent by circumstance, and 40 percent by factors within individuals' control, the result of choices they make.[14] This encouraging "happiness pie" concept went viral, leading to book contracts, speaking engagements, and other professional rewards for Lyubomirsky. Seligman transformed it into a "happiness formula" in his own work: $H = S + C + V$. That is, happiness, $H$, equals $S$ (genetic set point) plus $C$ (circumstances) plus $V$ (things under the individual's voluntary control). In part on the basis of Lyubomirsky's finding, he argued that there was a great deal of potential for the average person to become significantly happier.

Remarkably, it appears to have taken almost a decade and a half for anyone to critically evaluate Lyubomirsky's sunny claim in a peer-reviewed journal. But when Nick Brown and Julia Rohrer did, for an April 2019 article published in the *Journal of Happiness Studies*, they found many statistical problems. Among others, "there is only very limited evidence to place the figure for the heritability of well-being as low as (precisely) 50%. Consequently, there is little reason to believe that 40% is a reliable estimate of the variance in chronic happiness attributable to intentional activity—for example, if Lyubomirsky et al. had chosen a different (but, in our view, at least equally plausible) set of estimates, they might just as easily have concluded that as little as [5 percent] of variance in chronic happiness can be attributed to volitional activities."[15] Suffice it to say that there is a massive difference between 5 percent and 40 percent of an individual's level of happiness being within their control—a difference with obvious ramifications for the usefulness of positive psychology's books and interventions. Or, as Ruth Whippman put it in her book *America the Anxious*, "This is the 40 percent that anyone with a book to sell, a course of coaching to offer, or a happiness technique to market is hoping to co-opt."[16]

It's little surprise, in light of all this, that some external and internal critics have argued that positive psychology has made unwarranted claims. "The first data on rigorously tested positive psychology have only recently begun to show up in journals," wrote the positive psychologists Todd B. Kashdan and Michael F. Steger in 2011, "yet people have been offering to 'apply' positive psychology for several years already. What kind of message does this convey about the scientific endeavor of positive psychology? Is it any wonder that positive psychology is often dismissed as 'happiology' or the equivalent of accepting a Dixie cup of Kool-Aid from Jim Jones?"[17] As Horowitz writes in *Happier?*, a primary concern of some positive psychologists "involved the dangers of popularization," of polishing rough claims to make them look smoother for marketing purposes.[18]

But these serious questions surrounding the field's rigor haven't stopped positive psychology from quite successfully selling its wares to the public. It generates a great deal of press attention and millions in foundation support from donors like the John Templeton Foundation and has enjoyed many of the other perquisites of crossover success. Harvard's class on positive psychology was touted as the most popular in the university's centuries-long history.[19] Seligman's Positive Psychology Center, meanwhile, has been good, from both a public relations and a financial standpoint, for the university that hosts it. In its 2018 annual report, Seligman noted that "the PPC is financially self-sustaining and contributes substantial overhead to Penn."[20]

THE PPC HAS BEEN ABLE to build up a steady list of impressive clients over the years, ranging from police departments to medical schools, but its most important client is probably the U.S. Army. That's thanks to the fact that in 2008 the Army turned to Seligman to help it solve a crisis involving PTSD and suicide among soldiers, rewarding him and his academic home base with what would become many millions of dollars' worth of military contracts.

As we'll see, the Army's decision to contract with the PPC was understandable, at least from a certain angle. But to a keen observer of the Positive Psychology Center's offerings and promotional style, it might have come across as a questionable bet. On multiple occasions, Seligman and his center have made impressive claims about interventions that outpace the available evidence.

One example is the so-called Strath Haven Positive Psychology Curriculum intervention, which is named after the suburban Philadelphia high school where it was piloted. Its "major goals . . . are 1) to help students identify their signature character strengths and 2) to increase students' use of these strengths in day-to-day life," write Seligman and some of his colleagues in an article in the *Oxford Review of Education*. "In addition to these goals, the intervention strives to promote resilience, positive emotion and students' sense of meaning or purpose," they explain later.[21]

This approach is based mostly on ideas about the connection between character and well-being posited by Seligman and the late psychologist Christopher Peterson. The general claim is that certain aspects of positive character—bravery, honesty, and good humor, among others—are recognized across different cultures and that individuals' well-being and performance can be improved if they are taught to recognize and cultivate their own personal character strengths, which might be quite different from other people's.

In the *Oxford Review* paper, which summarizes multiple positive-psychology interventions, Seligman and his colleagues describe their efforts at Strath Haven as "the first empirical study of a Positive Psychology curriculum for adolescents. With a $2.8 million grant from the US Department of Education, our research group recently completed a randomised controlled evaluation of the high school positive psychology curriculum." The researchers explain that 347 ninth graders were assigned to either a class that included the curriculum ("approximately 20–25 80-minute sessions delivered over the 9th grade year") or a control group in which the school day was business as usual. "Students, their parents and teachers completed standard questionnaires before the programme, after the programme, and through two years of follow-up," write the authors. "Questionnaires measured students' strengths (e.g., love of learning, kindness), social skills, behavioural problems and enjoyment of school. In addition, we examined students' grades."[22]

Strikingly, even though these researchers were given almost $3 million to run a randomized controlled evaluation of their intervention, they never published one in full-blown, comprehensive form. The *Oxford Review* article's brief discussion of the program provides some provisional results, but few of the statistical details one would expect in a full-blown published evaluation. The authors make a somewhat vague claim about improved grades being seen in one subgroup (and presumably not others), but provide no statistics to back it up.[23] It's clear, though, that the general results were disappointing: "The positive psychology programme did not improve other outcomes we measured, such as students' reports of their depression and anxiety symptoms, character strengths, and participation in extracurricular activities."[24]

And yet if a school administrator curious about Seligman's track record visited the "Resilience Training for Educators" section of his Penn website, they'd find a rosier assessment: there, Seligman claims that the Strath Haven program "builds character strengths, relationships, and meaning, as well as raises positive emotion and reduces negative emotion." It's unclear why Seligman wrote that the program "did not improve . . . character strengths" in a peer-reviewed article and then that it "builds character strengths" on his academic website. When I asked the Positive Psychology Center if it could provide any further information about the evaluation process, I was referred to Jane Gillham, a Swarthmore psychologist unaffiliated with the PPC who helped design the program and who has worked with Seligman in the past (including as a co-author on the *Oxford Review of Education* article). She confirmed that a full-blown, peer-reviewed evaluation had never been published: "This was a research study conducted in collaboration with a local school district," she wrote. "The results were published in the Seligman et al. 2009 paper and then described in a couple chapters. The Seligman paper was peer reviewed, although you are correct that it isn't a typical empirical paper report."

Seligman's tendency to overclaim can also be seen in the case of the Penn Resilience Program, or PRP. PRP is one of the PPC's biggest "hits"; it has been purchased by schools all over the world and, as we will see, adapted to various noneducational settings as well.

PRP, which was first created by Gillham in the 1990s, is geared to ten- to fourteen-year-olds, and its goal is to improve these students' mental-health outcomes, particularly by making them more resistant to depression. As she and her colleagues explain in a book chapter, the program is based on Aaron T. Beck's model of depression, which "proposes that a tendency to view one's self, the world, and the future in overly negative ways, combined with a lack of behavioral coping skills, puts one at risk for depression and anxiety."[25] Much of what gives rise to depression, in this view, can be addressed via cognitive behavioral therapy.

We encountered CBT earlier: one key idea is to train the patient to adopt more adaptive, realistic ways of understanding and interpreting the world—to point them toward a better so-called explanatory style. As Gillham and her colleagues write, "An adolescent with a pessimistic explanatory style who fails a science test might think to herself, 'I'm stupid' or 'I have no ability in math or science.' If she gets an A on a science test, she might think, 'That test was easy.'"[26] If a young person's explanatory style can be improved, the thinking goes, it can bolster her sense of control over what happens to her ("I didn't do well on that test, but there are things I can do to improve next time, and it doesn't mean I'm hopelessly stupid"). And there is a solid base of evidence suggesting CBT is an effective treatment for anxiety and depression.

Usually, however, interventions like CBT are administered by a trained therapist to those who are already suffering from anxiety or depression. PRP, as it is often delivered, is quite different on both fronts. Remember that positive psychology, as an idea and an institution, is built around the belief that the field of psychology should seek to help people who are already healthy to thrive. So it goes with PRP: The goal is to instill, in healthy young people, cognitive habits and skills that will prevent depression and anxiety in the long run. It's generally meant to forestall rather than cure or manage mental-health problems (though there are also variants geared at groups of kids already exhibiting mental-health warning signs). One can see why this is such an appealing idea. If, in fact, there were some way to effectively inoculate young people, en masse, against these conditions, it would curtail a great deal of suffering (and save money in the process). The trainings themselves could also be cheaper than therapy or medication would be: PRP, unlike CBT (in most clinical settings), is not delivered by trained therapists. Rather, as Gillham and her colleagues write, PRP leaders "typically participate in a 4- or 5-day training workshop, where they first apply PRP skills in their own lives and then learn to deliver the curriculum to groups of late-elementary and middle school students" (here, too, there's some variation, since there are also versions of PRP delivered by mental-health experts).

The Penn Resilience Program, as laid out by Gillham and her colleagues, is delivered to groups of six to fifteen students over the course of about twenty hours, total, though the number and length of individual sessions can vary. The primary purpose is to help the children and early adolescents better understand basic cognitive behavioral principles, including the potential harms of negative self-talk (*I failed this test; I really am just worthless*) and catastrophization (*My mom was supposed to be home by now; she must have gotten into a horrible accident*).

PRP teaches students to identify these patterns of rigid, inflexible thinking and offers them tools for snapping out of them. At one point, for example, the curriculum uses "the detective story" to help instill these messages: "The first detective (Merlock Worms) thinks of one possible suspect and is certain that this person committed the crime because 'his was the first name that popped into my head.' The second detective (Sherlock Holmes) makes a list of suspects (analogous to generating alternatives) and then carefully examines the evidence for and against each one."[27] PRP also presents students with specific techniques for improving their mental health, like following up a worst-case explanation that pops into their head with a best-case one (*Maybe my mom won the lottery and* that's *why she's late*) and then recognizing that *both* are quite unlikely. If kids can be trained to do this, the thinking goes, they will develop more flexible, adaptive behavioral and cognitive habits, buttressing their mental health in the long run.

The Positive Psychology Center clearly views the Penn Resilience Program as one of its premier offerings, which is understandable given how many parents and teachers and school administrators are concerned about the apparent (though not empirically uncontested) rise in depression and anxiety among young people.[28] A web page devoted to the program boasts that it increases "well-being and optimism" and lowers "depression, anxiety, and conduct problems, [as well as] substance abuse and mental health diagnoses," in addition to improving physical health.[29] In a talk he gave at the 2009 annual conference of the American Psychological Association, Seligman presented the results of a review of nineteen PRP studies conducted over twenty years. The accompanying APA press release noted that "based on the students' assessments of their own feelings, the researchers found that PRP increased optimism and reduced depressive symptoms for up to a year. The program also reduced hopelessness and clinical levels of depression and anxiety. Additionally, the PRP worked equally well for children from different racial/ethnic backgrounds."[30]

Unfortunately, Seligman doesn't appear to have ever published this review of the literature anywhere (I did ask him about this directly in an email, and in response he pointed me to other research instead), so it's unclear how impressive the effects he found were, what criteria he used to include or exclude given studies, and so forth. But another, more formally conducted review of the literature—a meta-analysis co-authored by Gillham herself—came to a different conclusion.

To be sure, meta-analysis is no panacea. If an entire body of research suffers from certain types of biases, which can occur, then any meta-analyses drawn from it will be similarly flawed. And there are often fierce debates about which studies should and should not be included in a meta-analysis, so there's plenty of fallible human judgment involved in this type of research, too. But in most contexts, a rigorously conducted meta-analysis *is* preferable to a handful of studies picked without clearly defined inclusion criteria or quality-control standards in mind.

The meta-analysis Gillham co-authored, lead-authored by Steven Brunwasser and published in 2009, examined seventeen controlled evaluations of the PRP—that is, studies that compared the outcomes of a PRP group and a control group. It found that while the PRP did appear to reduce depressive symptoms among students exposed to it, those reductions were small, statistically speaking. "Future PRP research should examine whether PRP's effects on depressive symptoms lead to clinically meaningful benefits for its participants, whether the program is cost-effective, . . . and whether PRP is effective when delivered under real-world conditions," the authors concluded.[31]

This is not an impressive evaluation, referring as it did to a program already being sold to schools on the basis of its supposedly impressive evidence base. And in 2016, the *Journal of Adolescence* delivered an even harsher verdict about PRP in another meta-analysis. "No evidence of PRP in reducing depression or anxiety and improving explanatory style was found," the authors wrote. "The large scale roll-out of PRP cannot be recommended. The content and structure of universal PRP should be reconsidered."[32] Gillham said in an email that she disagreed with that meta-analysis's inclusion criteria, because she believes it lumped in studies of programs that were fairly different from PRP. But the point is that neither meta-analysis delivered a true vote of confidence for PRP.

Seligman, for his part, pointed me to a 2015 meta-analysis conducted by researchers in Australia and New Zealand, which appears to show that PRP has some effectiveness. But, while I'll relegate the details to a footnote, a close look reveals that that meta-analysis doesn't tell an appreciably different story from the one told by Brunwasser and Gillham's evaluation, especially when it comes to the specific "flavor" of PRP that makes the program so attractive on the grounds of potential cost-effectiveness—one in which the intervention is delivered to (mostly) healthy students by laypeople who can be quickly trained for that task. In fact, the co-authors themselves write that "Our results are consistent with another review of the PRP"—and cite the underwhelming meta-analysis by Brunwasser and Gillham.[33]

Why isn't PRP more effective? While no one knows for sure, it's possible to engage in some informed speculation by recognizing the context dependency of psychological research: because humans are so complicated and act (or react) so differently in different settings, a discovery that holds among one group in one time and place might not apply more generally. (This is one of the reasons there is widespread and growing concern that so many past psychology studies have been conducted exclusively on so-called WEIRD subjects—that is, Western, Educated people from Industrialized, Rich, Democratic countries. There is well-justified suspicion that findings applicable to WEIRD subjects may not generalize to others, that "We Aren't the World," as the headline of an important article in *Pacific Standard* about the anthropologist Joe Henrich's research on this subject puts it.[34])

The evidence base for cognitive behavioral therapy consists generally of studies involving one-on-one therapy with individuals who are already suffering from depression or anxiety. It would be beneficial to society (not to mention to the Positive Psychology Center) if it were possible to generalize this finding from an individual treatment program to a *group prevention* program. But because interventions that work in one context might not work in another, this was always a theory to be tested, not an obviously true claim.

The difference between one-on-one CBT and the PRP is significant: in a PRP context, unlike during traditional CBT, the individuals receiving the intervention (mostly) don't already have depression or anxiety, and they're not receiving much individualized attention. Think of the difference between a fourteen-year-old who already has a specific set of mental-health symptoms going to a cognitive behavioral specialist twice a week for forty-five-minute sessions, enjoying the full, undivided attention of a therapist who has trained for years, and an otherwise similar fourteen-year-old, not mentally ill, sitting through twenty hours of group exercises led by a non-therapist with only four or five days of training. It's a radically different context. For what it's worth, Gillham herself agreed that these two variables probably explain the underwhelming results of the meta-analysis she co-authored. "I think you are probably correct these are likely to be huge factors," she wrote in an email.

None of which is to say the basic overall principles of PRP are unsound; CBT is held in high esteem by many clinical psychologists as a means of treating depression and anxiety (though there are some skeptics who believe the evidence for it is overblown). The program rests on ideas that are more solid than many others associated with the positive psychology movement, like the critical happiness ratio or the happiness pie. But PRP *was* a novel and experimental approach to forestalling depression and anxiety—one that demanded evidence—and it doesn't appear to have delivered on its early promise.

Whatever the reasons for PRP's shortcomings, as of 2020 many of the Positive Psychology Center's clients around the world don't appear to have gotten the message; they keep purchasing it. In fact the program's purview has expanded beyond schools. The PPC's 2018 annual report touts the fact that the center received a two-year grant from the Department of Justice to adapt PRP for law-enforcement personnel, as well as contracts to develop similar programs for the medical schools at Yale and Penn, among myriad other clients. Overall, notes Seligman in his report, "Since 2007, we have delivered more than 270 Penn Resilience Programs to more than 50,000 people." The fact that these are adult contexts adds a whole other layer of uncertainty given that PRP was designed for kids.[35]

Why does the lack of evidence seem to matter so little here? Recall the psychologist Matthew Grawitch's explanation of how unskilled intuition works at the institutional level. The Penn Resilience Program certainly provides a "plausible explanation for how its intervention/approach will fix the problem," and one that happens to be based, at core, on some fairly widely accepted science: *We can head off youth mental-health problems before they even occur, simply by delivering twenty hours or so of training based mostly on cognitive behavioral principles to groups of students.* It's easy to understand how this story might sound perfectly reasonable to a school administrator who is an expert at, well, school administration but not sufficiently well versed in adolescent psychology or the ins and outs of CBT research or research methodology to ask the right questions. The problem is that stories that sound good don't always turn out to be true.

THE POSITIVE PSYCHOLOGY CENTER has had a substantial impact on education around the country, but the adoption of its ideas by the U.S. military may be more consequential still. It is here where what Daniel Horowitz calls the "dangers of popularization" come most clearly into focus.

Around 2007 the U.S. Army realized it had a full-blown mental-health crisis on its hands. The wars in Iraq and Afghanistan had stretched personnel thin, to the point where, in order to keep numbers up, the Pentagon was forced to revise long-standing rules about the length of combat deployments.[36] This led to more and longer deployments, with less time off between them, and to increasing numbers of National Guard and U.S. Army Reserves personnel—individuals who had, in many cases, signed up imagining their service would entail "one weekend a month, two weeks a year" of peacetime drills and perhaps occasional domestic disaster relief, as one slogan for the National Guard put it—being sent repeatedly to active combat zones.[37]

Soldiers in both Iraq and Afghanistan, like their predecessors in Vietnam, became occupiers of lands where they mostly didn't speak the language, often couldn't tell friend from foe, and were beset by threats that came seemingly out of nowhere—sudden ambushes and explosive traps and other deviations from the forms of conventional warfare where American military forces enjoy their most obvious advantages. The results were staggering: about 15.7 percent of deployed veterans and 10.9 percent of non-deployed veterans screened positive for PTSD during this era, according to a major study,[38] compared with a lifetime prevalence of about 6.8 percent in the general population.[39] In 2002, a terrifying uptick in suicides among Army soldiers, who bore the brunt of the conflicts, began, and many of those deaths appeared to be directly connected to PTSD symptoms.

PTSD is a difficult disorder that seems almost designed, in certain ways, to stymie treatment. At root, experts believe it is the result of a shock to the nervous system causing certain aspects of its fight-or-flight response to misfire. In many cases, the body subsequently reacts to everyday situations as though they were life-and-death encounters. PTSD victims often suffer from debilitating flashbacks, replete with intense physical symptoms like a racing heart, when they're exposed to otherwise innocuous stimuli with some connection to their trauma: if they suffered an ambush on a bridge, they may find it almost impossible to come anywhere near a bridge without enduring a horrific physical reaction that is very hard to explain to anyone who lacks direct experience with the disorder. Those with PTSD also suffer from "intense, disturbing thoughts and feelings related to their experience that last long after the traumatic event has ended," the American Psychiatric Association explains, and their symptoms can include "flashbacks or nightmares," anger, fear, and a sense that they are "detached or estranged from other people."

The site also notes that "people with PTSD may avoid situations or people that remind them of the traumatic event, and they may have strong negative reactions to something as ordinary as a loud noise or an accidental touch."[40] *Avoidance* is one of the most insidious aspects of PTSD because of how it locks the victim away: as a general rule, if you are suffering psychologically and feel separated from other people, from social support and exposure to normal daily routines, your situation is unlikely to improve and may well worsen. "Trapped in your own head" may be a cliché, but it accurately describes the worst cases of PTSD. In the accounts of those who have suffered from it, the themes of repetition, exhaustion, and being held hostage by one's own body often recur. Jason Kander, a former Democratic secretary of state of Missouri who served in Afghanistan and suffered from PTSD, explained to the journalist Chris Hayes the toll "becoming hypervigilant" took on him in a 2019 interview, reflecting on the compulsive way he was "always thinking about how many exits there were [and] having four different plans as to how you were going to get out of a situation"—a "situation" in this case meaning something like a night out at a restaurant. "I went about twelve years without a good night's sleep," he told Hayes.[41]

Given how many soldiers were fighting in Iraq and Afghanistan, the brutal nature of the combat many of them experienced, and the link between PTSD and potentially deadly behaviors like addiction and violence, it's no surprise that the epidemic of untreated PTSD sparked by the post-9/11 wars led to an endless stream of terrible tragedies back home. While suicide is a far more statistically likely outcome for those afflicted by PTSD, it was a spate of senseless murders committed by veterans that truly seized the nation's attention. "Town by town across the United States, headlines have been telling similar stories," reported *The New York Times* in early 2008.[42] "Lakewood, Washington: 'Family Blames Iraq After Son Kills Wife.' Pierre, South Dakota: 'Soldier Charged With Murder Testifies About Postwar Stress.' Colorado Springs: 'Iraq War Vets Suspected in Two Slayings, Crime Ring.'"[43] (The Army is overwhelmingly male, but female soldiers fighting in Iraq and Afghanistan were disproportionately affected by another issue: sexual assault. According to Veterans Administration statistics, one in four women seen at VA clinics, and one in a hundred men, report having experienced sexual assault during their service, and military sexual assault is "commonly associated" with PTSD.)

In October 2007, Colonel Jill Chambers, an energetic survivor of the 9/11 attack on the Pentagon and herself a PTSD sufferer, was handed the monumental task of figuring out how to solve this problem. Admiral Michael Mullen, who had just arrived at the Pentagon as the new chair of the Joint Chiefs of Staff, named her "Special Assistant to the Chairman for Returning Warrior Issues" and gave her a simple imperative, as she described it to me: "Jill, go forth, get away from the Pentagon—get out there and start talking to people and find out what it is that's causing our service members so many problems."

Chambers took on her new role with gusto. "For the next eight months, I was out and about all over the world," she explained. She had contacts throughout the armed forces, so she traveled all over the country to have conversations with those who were shouldering the heaviest load of the ongoing wars. "It got to be, *Jill's coming in, she's cool, please get your guys to talk with her,*" Chambers explained with evident pride. Over and over, soldiers back from Iraq and Afghanistan would tell Chambers stories of trauma tinged with stigma.

One said he had been sleeping in his garage because he kept waking up to find himself choking his terrified wife. He was scared of his own behavior, but also scared of speaking about it aloud. "Look, if you tell anybody about this, I'm going to deny it," he told Chambers.

This was a crisis, and it was clear the military needed to do something. That *something* arrived via a coincidence. One day, Chambers was on a flight from Washington, D.C., to Boston with her husband, the country musician Michael Peterson, and he nudged her. He was reading a book called *Learned Optimism: How to Change Your Mind and Your Life*, written by a psychologist named Martin Seligman. It seemed relevant to Chambers's current work; Peterson's key takeaway from *Learned Optimism*, as Chambers explained it, was that "you can really prime your pump before you face adversity to actually get yourself prepared for it." Could there be a way to instill in soldiers a sense of resilience and optimism that would help them both during combat and after, that would effectively inoculate them against the worst psychological ravages of war? "Why don't you just call Marty Seligman?" Peterson asked Chambers. So she did.

In August 2008, Chambers and Peterson met with Seligman in the garden of his Philadelphia home and came away very impressed. A few more calls and meetings later—and some pushback from Army higher-ups who wanted to sweep the problem under the rug but whom Chambers could brush off because of her direct mandate from Mullen ("Four stars beat any of those two- and three-star generals," as she put it)—and Seligman had earned a meeting with General George Casey, chief of staff of the Army, to whom Mullen had delegated the task of vetting him.

Casey proved a quick convert. "He put his fist down and he said, *By golly, we have a problem, and we are going to start talking about post-traumatic stress*," Chambers recalled. Seligman, armed as he was with what appeared to be reams of research and impressively rigorous books supporting his approach, emerged from the meeting as the Army's go-to guy for addressing this newly acknowledged crisis. "Who else out there had a resilience-building program, right?" said Carl Castro, a retired colonel who was involved in multiple Army mental-health initiatives, including Comprehensive Soldier Fitness, and who is currently a social-work professor at the University of Southern California. "Who else had a validated program, some data, any data around building resilience? And if you go back and look in the literature, there was only one person."

And that's how the CSF program was born. Soon, it was a mandatory part of Army life for every soldier: more than a million in all. It would become one of the largest mental-health interventions geared at a single population in the history of humanity, and possibly the most expensive.

COMPREHENSIVE SOLDIER FITNESS is a hybrid consisting of three different components: One is a set of online-learning modules geared at boosting mental health adapted from Battlemind, a preexisting military mental-health program. Another is a mandatory annual survey, the Global Assessment Tool, or GAT, which was cobbled together from a number of different instruments. All soldiers are required to take the GAT every year and to complete a set number of hours of the online modules.

But the centerpiece of CSF, at least when it came to how it was advertised to the public, was the Master Resilience Training program. MRT, as it is known, is a train-the-trainer program closely modeled after the PRP, and like the online-learning modules and the GAT, it is mandatory for all soldiers. It was also the main reason Seligman's Positive Psychology Center won its initial $31 million from the Army, in the form of a 2010 no-bid contract, as well as the funding that would follow.[44] That initial contract contained common Pentagon budgeting language indicating that the recipient is the only provider of a particular service: "There is only one responsible source due to a unique capability provided, and no other supplies or services will satisfy agency requirements." It's clear, from Seligman's account of the early days of CSF in his book *Flourish*, as well as various statements from others, that the military's claims about "unique capability" stemmed from the supposedly strong evidence base for PRP.[45]

Seligman's argument was that Comprehensive Soldier Fitness could help reduce PTSD and suicidality, and it's worth pausing here to reflect on how many steps removed this claim is from the initial goals and scope of the Penn Resilience Program. When the PRP began, it was novel and untested, given that it was attempting to *prevent* depression and anxiety with tools that had only been validated for *treating* those conditions. But still, it was premised on a reasonable theory in light of CBT's solid base of evidence, and all it was claiming was that it could prevent anxiety and depression in some students.

Comprehensive Soldier Fitness was founded on a more radical claim: that an adapted version of the PRP could prevent PTSD and therefore suicide. "That's why we instituted the Comprehensive Soldier Fitness Program," General Casey told a Senate Appropriations subcommittee in 2010, "to give the soldiers and family members and civilians the skills they need on the front end to be more resilient and to stay away from suicide to begin with. It's a long-term program, but I think that is the only way that we are ultimately going to begin to reduce this."[46]

PRP itself, though, was never designed for anything remotely like that; no one associated with it, until Seligman linked up with the Army, appears to have ever claimed it could prevent PTSD or suicide, and such an idea wasn't even on the radar of the program's designers as they built it. And yet in one paper Seligman, his Penn positive psychology peer Karen Reivich, and Sharon McBride of the Army wrote of the Penn Resilience Program—which by that time had been shown in a meta-analysis not to be particularly effective in reducing depression among ten- to fourteen-year-olds—that "the preventive effects of the PRP on depression and anxiety are relevant to one of the aims of the MRT course, preventing posttraumatic stress disorder (PTSD), since PTSD is a nasty combination of depressive and anxiety symptoms."[47]

But just because PTSD can cause depression or anxiety doesn't mean that treating depression and anxiety cures PTSD, or that preventing depression and anxiety prevents PTSD. Having a cold might make you cough, but simply curing the cough may fail to address the underlying illness.

As the researcher and data sleuth Nick Brown (who helped debunk the happiness pie and critical positivity ratio concepts) wrote in a critical review of Comprehensive Soldier Fitness published in the open-access online academic journal *The Winnower*, "Much of PTSD consists of symptoms whose prevention is not addressed by the PRP, or indeed anything else that comes under the umbrella of positive psychology." He delivered an unflinchingly harsh verdict about the chain of causal claims Seligman had sold to the Army: "The idea that techniques that have demonstrated, at best, marginal effects in reducing depressive symptoms in school-age children could also prevent the onset of a condition that is associated with some of the most extreme situations with which humans can be confronted is a remarkable one that does not seem to be backed up by empirical evidence."[48]

Stretching things even further, Seligman and his colleagues didn't merely adapt PRP to a new and unfamiliar context; they also bolstered the Master Resilience Training with components taken from other corners of positive psychology, many of them involving attempts to make people a bit more optimistic in general. "Resilient people bounce, not break," reads one slide from an MRT session. Under that, two images: "You" over a tennis ball, "Not you" over a cracked egg with yolk oozing out. A bit later in that same slide deck, an in-class exercise: "Discuss resilience using the quotes [from earlier slides], your personal experiences, and what we've discussed so far in the course. Create a list of the strengths, skills, and abilities that you believe are critical for resilience." Later still, the module promises that resilience "can be developed: Everyone can enhance his or her resilience by developing the MRT competencies." Elsewhere, soldiers are instructed to "hunt the good stuff"—that is, to remain optimistic by thinking of the good things in life. (This became somewhat infamous among some of the critics of Comprehensive Soldier Fitness with whom I spoke: before sending a twenty-year-old into an urban-combat hellhole, you're reminding him to "hunt the good stuff.")

Other aspects of positive psychology were shoehorned into the CSF curriculum as well, such as the aforementioned theory about finding and cultivating character strengths, which does not appear to have ever been tested as an anti-PTSD or anti-suicide measure in any context. These materials were not from PRP as it was originally conceived, and they have a far weaker evidence base than interventions premised on cognitive behavioral principles. In an email Gillham, after cautioning that she was unfamiliar with CSF itself, noted that "the original PRP did not include positive psychology activities. I personally don't consider the original PRP a positive psychology intervention."

Overall, there was no evidence PRP itself could prevent PTSD or suicide in its existing form; Seligman and his colleagues then padded it with elements that are, according to the available evidence (or lack thereof), even less suited to that task.

WHEN UNSKILLED INTUITION COLLIDES with slickly advertised behavioral science, the problem isn't just that institutions may end up spending money on approaches that don't work; they may also end up neglecting approaches that do work. Hype and excitement and overclaiming can misdirect an institution's priorities in damaging ways.

Contra the claims of Seligman and his colleagues, effective treatment of PTSD is far more complicated than addressing the anxiety and depression symptoms it can cause, or "building resilience" in some general sense. Preventing PTSD, as the Army sought to do, is even harder terrain. In fact, at the time CSF was launched, there was little evidence *any* intervention could prevent PTSD, and even as I write this, midway through 2020, researchers are taking only their first steps toward pilot programs that might be able to accomplish that goal.

But there are effective treatments for PTSD. One of them is called cognitive processing therapy, and it was developed by Patricia Resick of Duke University, a leading trauma researcher. CPT is itself premised on a cognitive behavioral approach, but it is both more targeted and more intense than anything positive psychology attempts: it helps patients with PTSD rewire how they conceive of the traumas and tragedies they have experienced. A soldier might be hung up on the idea that he did something wrong, leading to a buddy's death, for example. So a therapist might in turn help gently lead him to better understand that war really is chaotic, random, and unfair, and that therefore he shouldn't hold himself responsible for such a horrible event. CPT is focused on ameliorating cognitive distortions, but, crucially, it can do so only among those who are already afflicted with PTSD, because the approach depends upon untangling the *specific* thought patterns and experiences of a given sufferer.

CPT is considered a "gold standard" PTSD treatment by the Pentagon. Another such treatment is prolonged exposure, or PE, therapy.[49] Developed by the Israeli researcher Edna Foa, who is based (as fate would have it) at the University of Pennsylvania's Perelman School of Medicine, PE therapy entails helping patients to face down and process their trauma and its triggers, rather than fall victim to the avoidance strategies that so often cut them off from other people and stymie their ability to integrate their trauma into a recuperated sense of self. Neither of these treatments is perfect, and both have their critics. But the available research strongly suggests that the average veteran with PTSD would benefit from a course of PE therapy or CPT.

The problem, though, is that the military has long had a serious problem getting veterans to enter and stick with these treatments.

The numbers are stark: a 2017 paper found that only 56 percent of returned veterans from Iraq and Afghanistan who screened positive for PTSD had *any* subsequent engagement with mental-health services, and over the years the rates at which veterans with PTSD have partaken of therapies like PE or CPT have hovered at just about a third.[50] Research suggests that veterans' obstacles to treatment range from logistics—some are simply unable to get to a VA center or other treatment facility on a regular basis—to stigma against open discussion of trauma symptoms.[51]

So at the time the U.S. military faced its burgeoning PTSD crisis, there was one rather obvious approach to take: expanding access to scientifically validated treatment for veterans, and seeking to better understand why they often shied from or felt cut off from it. But that wasn't what happened.

In his excellent book about human reason giving and storytelling, *Why? What Happens When People Give Reasons . . . and Why*, the late sociologist Charles Tilly offers a wonderfully pithy and sympathetic account of what it's like to work in an organization and why questionable stories can catch on at the expense of more rigorous ones in those settings: "So many demands are competing for our attention! But that is the point: which demands, which reasons, which relationships, and therefore which sorts of reliable information reach us depend on historically established organizational routines over which we exercise limited control."[52]

The Army's leadership appears to have been particularly drawn to PTSD interventions that would piggyback on its institutional fixation with "resilience" and prevention. It's understandable why: First, the prospect of preventing PTSD rather than having to treat it after the fact was likely irresistible to many who understood how bad the situation had gotten. "Build resilience, prevent PTSD" was too good a promise to refuse, because if kept it would forestall a tremendous amount of human suffering, and to the unskilled layperson there *did* appear to be evidence supporting this approach, in the form of Seligman's impressive claims about PRP. But CSF also fit neatly with the Army's beliefs in self-possession and self-efficacy, meaning that it could be pitched to Army bigwigs in a language they were already fluent in. (A salesman who understands the internal culture of a company, or who already inhabits it himself, is going to have a better time selling that company something.) On top of all that, for the Army to introduce a sweeping new program rather than bolstering or tweaking preexisting ones would bring with it obvious PR opportunities, such as videos of soldiers participating in an exciting, novel mental-health initiative (which were indeed shot and disseminated to the public).

The adoption of Comprehensive Soldier Fitness was not driven by PTSD experts. Neither the Pentagon staffer initially tasked with narrowing down the Army's range of potential options for addressing the PTSD crisis (Chambers) nor the general who became CSF's arguably fiercest advocate (Casey) was an actual expert on PTSD, nor was Seligman himself. But, to once more borrow Matthew Grawitch's language about unskilled intuition, the Army personnel who were convinced "Marty" was their man encountered a "plausible explanation" for how CSF would "fix the problem" of PTSD—one couched in "a good story" (resilience boosting). None of this is particularly surprising in light of how human institutions work.

That doesn't mean the Army should be let off the hook. While some PTSD experts weren't looped in at all—"We were never asked to consult on prevention of PTSD, or whether this program would work, or whether it should be funded," said Resick—there were skeptics involved in the process, but it didn't seem to matter. "No one was happy with the level of evidence with Marty's program, by the way," said Carl Castro, the retired colonel heavily involved in the development and rollout of both CSF and other Army mental-health programs. "Everyone recognized that there were significant shortcomings in the data that existed for Marty's work, but it was the best we had." Simply put, many experts in the relevant areas did not find the CSF storyline credible. "When I first heard about it I was more or less floored," George Bonanno, a clinical psychologist at Columbia University and leading resilience researcher, told the journalist and American studies scholar Daniel DeFraia for a 2019 article he published in *The War Horse*, a military-focused journalism outlet. "I've been studying resilience for 20 years, and I don't know of any empirical data that shows how to build resilience in anybody."[53]

Richard McNally, a Harvard psychologist and leading PTSD expert, told me he was invited to an early meeting with Seligman, General Rhonda Cornum (another major figure in the development of CSF), and others at the Positive Psychology Center to discuss the nascent program. There, he tried to emphasize just how little evidence there was that the adapted-PRP approach would successfully address PTSD. "That's why my suggestion to Marty and to General Cornum was 'Why don't you pilot this first, and then you can tweak it, improve it, et cetera, et cetera, until you can get a good sense of whether this is going to work prior to disseminating it throughout the entire Army?'" he recalled. "That was my issue, because there was not a great deal of evidence on this."

With pilot testing, the idea is simply to roll out a smaller-scale version of the program in question on a subset of the population for which it's designed—not only to test for evidence of its efficacy, but also to ensure it has no adverse effects. But General Casey, besotted as he was with Seligman's ideas, would have none of this pilot-test talk. During a key exchange recounted in *Flourish*, Seligman described what happened when he and Cornum asked Casey for an initial pilot test to see how their program performed. "Hold on," the general "thundered," as Seligman put it. "I don't want a pilot study. We've studied Marty's work. They've published more than a dozen replications. We are satisfied with it, and we are ready to bet it will prevent depression, anxiety, and PTSD. This is not an academic exercise, and I don't want another study. This is war. General [Cornum], I want you to roll this out to the whole Army."[54]

This is a veritable carnival of unskilled intuition and exaggerated storytelling, a striking example of how science can be adulterated and misunderstood by an organization seeking to apply it. "They've published more than a dozen replications": well, but when those and other studies were meta-analyzed, PRP didn't seem to *do* much, and those studies were conducted on kids, anyway. "We are satisfied with it, and we are ready to bet it will prevent depression, anxiety, and PTSD": no published literature on PRP claimed it could prevent PTSD, because that wasn't what it was designed for.

Unsurprisingly, the Army never produced any real evidence CSF works. While it did publish four "technical papers," none of them peer-reviewed, that purported to show the effectiveness of CSF ("Study concludes Master Resilience Training effective," touted the Army's website[55]), these analyses don't survive close scrutiny. The psychologists Roy Eidelson and Stephen Soldz, for example, published a working paper showing that the Army's evaluations were riddled with cherry-picking and basic methodological errors; in one instance, for example, the outcomes for soldiers who hadn't been deployed were compared with those for soldiers who had, introducing a mega-confound that renders the comparison meaningless, because it would be impossible to know whether any differences between the groups should be attributed to CSF or to deployment itself.[56]

But it's the Global Assessment Tool—the mandatory survey soldiers have to fill out every year—that most fully exposes the hasty, haphazard way in which Comprehensive Soldier Fitness was developed and implemented. The goal of the GAT was to collect ongoing data on soldiers' well-being and to allow the Army to measure whether CSF was doing what it was designed to do. When Seligman told the APA's *Monitor on Psychology* magazine in 2009 that CSF "is the largest study—1.1 million soldiers—psychology has ever been involved in, and it will yield definitive data about whether or not [resiliency and psychological fitness training] works," he was referring to the GAT and the data it would supposedly produce.[57] (One can't help but wonder if Seligman hoped that decades from now people would look back on the GAT the way we look back on the IQ test, put to its most important early use by the Army, as a tool for sorting recruits, during World War I.)

The problem, as Eidelson and Soldz explained, is that "the GAT does *not* include any validated [that is, previously tested and shown to be accurate] measures that assess PTSD, depression, suicidality, or other major psychological disorders, even though preventing these disorders is a key goal of the CSF program."[58] In his Army War College master's thesis, Colonel Richard Franklin Timmons II, citing Eidelson and Soldz's research, highlighted the same problem,[59] as did Nick Brown in his *Winnower* article. Instead, the psychological portion of the GAT (which also measures other characteristics, like soldiers' level of physical fitness and strength of family ties) consists largely of instruments adapted from positive psychology, particularly regarding character strengths. Some of the questions, for example, focus on the test takers' perception of whether and to what extent they have exhibited various character strengths—among them "bravery or courage," "honesty," or "zest and enthusiasm"—in the previous four weeks, as measured on an 11-point scale. But because survey items like this have little empirical evidence behind them (compared with validated instruments for measuring depression or suicidality), it's impossible to know how to interpret it when, for example, a soldier's level of self-assessed "zest and enthusiasm" ticks up a couple points from one year to the next. So the main instrument the Army built to measure CSF's effectiveness in preventing PTSD and suicide was wholly incapable of performing this task; it was impossible for the Army to generate any useful data about the metrics it was most interested in.*

---

* For the sake of thoroughness, and to assist readers hoping to learn more about Comprehensive Soldier Fitness online, I should note that the program has progressed through multiple versions over the years. For example, Comprehensive Soldier

The Institute of Medicine, an august branch of the National Academy of Sciences, came to the same conclusion as Eidelson, Soldz, Brown, and the program's other critics in a major 2014 report that evaluated the military's various efforts for improving the psychological well-being of service members and their families: "Although evaluations that were conducted by CSF staff and were not subject to peer review have demonstrated statistically significant improvement in some GAT subscale scores, the effect sizes have been very small, with no clinically meaningful differences in pre- and posttest scores. Accordingly, it is difficult to argue there has been any meaningful change in GAT scores as a result of participation." In addition, the IoM report notes, the one attempt the Army made to evaluate CSF on the basis of actual diagnoses among service members found "no difference in diagnosis among those receiving the [CSF] intervention" and those who had not participated in it.[60]

None of this was cheap. There's some fuzziness to the numbers, but in 2017 the Army told Daniel DeFraia that CSF cost $43.7 million the previous year. This tracks, roughly speaking, with the *USA Today* journalist Gregg Zoroya's estimate that as of 2015 the program had been a six-year, $287 million enterprise (like DeFraia, Zoroya is one of the few journalists who has dug deeply into the program).[61] Of course from a military perspective this is peanuts: a single F-35 costs about $80 million. But if, as the numbers suggest, CSF has cost the

---

Fitness (CSF) became Comprehensive Soldier and Family Fitness (CSF2), which as the name suggests includes programming for soldiers' family members. The GAT was supplanted by the GAT 2.0, which in turn was renamed the Azimuth Check—its present name as of August 2020. (I've stuck with "GAT" throughout this chapter because that has been the instrument's name for most of the existence of CSF, and, to be honest, because it rolls off the tongue a bit better.)

But there's been sufficient continuity to the program that the descriptions and critiques I offer in this chapter apply to its multiple iterations. For example, a source sent me the full versions of both the GAT 2.0 and the Azimuth check, and all the psychologically-oriented questions on the Azimuth Check come straight from its predecessor. As for the program's other content, as of August 2020, an army spokesman told me that content adapted from the Penn Resilience Program is still part of CSF.

Army somewhere in the neighborhood of half a billion dollars since it was launched more than a decade ago (and has cost the Pentagon even more if you include the funds spent on the Air Force version, Comprehensive Airman Fitness, which launched in 2011[62])—that's still a tremendous amount of money, absolutely speaking, when one considers the good it could do in helping get soldiers the mental-health care they need. There may be no other single mental-health intervention in the history of humanity that has cost this much, and the Army has almost nothing to show for it.

TREATING MILITARY PTSD ISN'T pretty or easy or photogenic. Many of Patricia Resick's patients have been part of "seven, eight, nine deployments" and have endured trauma most of us couldn't imagine: "They're seeing their friends being blown up, they're picking up body parts, they're nearly being killed themselves, they're having post-concussive syndrome," she told me. And by the time they get to Resick, they have often internalized certain ideas about the nature of war and agency and responsibility that are very important to the military, but directly counterproductive to the goal of recovering from trauma. A common task with her patients, she said, is "undo[ing] the learning that they learned while they were in the military. And it's pretty hard, because a lot of my studies are with active duty and they're still being what I would think of as brainwashed."

For example, Resick explained that as they are trained for combat, soldiers are often given some version of the story line *If everybody does their jobs correctly, everybody will come home*. But reality's messiness often defeats military mottoes, and in real-life combat that one's plainly false. "And so everybody didn't come home," Resick explained, "and now they look to themselves and they look to the person next to them and say, 'What did we do wrong? Somebody must not have done their job or we would have all come home.' That's ludicrous. I mean, you plant a mine deep enough nobody can see it. It has nothing to do with how well you did your job . . . If you're taught over and over and over and over again, and you're only nineteen years old, and your brain hasn't finished developing, that if everybody does their jobs okay then everybody's coming home, then they feel guilt." Soldiers tend to be young, so "their executive functioning in their brains hasn't even finished developing yet, they get these rigid kinds of ideas, and we've got to spend our time kind of undoing their learning and helping them really look at the context of the event and how there wasn't anything they could do at the time. The buried IED was not to be seen; how many other people were there who didn't see it? Why is all this blame falling on you?"

Resick's insight is as fascinating as it is heartbreaking and brings to mind the World War II veteran and performance artist Audie Murphy's line about postwar treatment of veterans: "After the war, they took Army dogs and rehabilitated them for civilian life. But they turned soldiers into civilians immediately, and let 'em sink or swim."[63] To actually treat PTSD often entails deprogramming the military's very own messages. You need to make people realize they did *not* have control over situations that were in fact chaotic and violent and incomprehensible. The training that was designed to help make individuals better fighters, and to improve the chance that they would get out of these situations alive, can be a roadblock to their full recovery from trauma. Sometimes the cognitive distortions are coming from inside the house.

I keep going back to Jill Chambers, a kind, energetic, intelligent woman who genuinely wanted to help soldiers. I keep going back to all those interviews she conducted with experts and with soldiers. I keep going back to the question of whom she was most likely to listen to, and why: back to unskilled intuition and the very Army-specific ways it manifested itself in this story. Seligman and positive psychology speak the same language as the Army. The Army was made to believe it could simply double down on the same messages it had always sought to impart—messages about toughness, resilience, and optimism—and, in doing so, ameliorate a PTSD crisis. The simple, inspiring mottoes that define military culture embody a great deal of what Seligman and his field of positive psychology stand for: individuals have significant potential to improve their situation and their probability of success, so long as they embrace hard work, dedication, and an optimistic attitude.

There is nothing wrong with telling people to be optimistic and hardworking and dedicated, of course. It's a truism that all else being equal, a well-prepared person will fare better than an ill-prepared one. But wartime PTSD is one of many complicated human problems that doesn't quite obey this sort of logic, at least not neatly. There's little evidence teaching people resiliency or optimism skills can shield them from PTSD.

The absence of evidence that the Penn Resilience Program and Comprehensive Soldier Fitness actually work as anti-PTSD interventions—as well as evidence that other approaches *do*, on average, work (at least as far as treating trauma that has already been inflicted)—was right there in the literature all along. Patricia Resick and Richard McNally and Edna Foa could have told anyone who asked, and in some cases *did* tell those who asked, that what the Army was rolling out was based on no one's expert understanding of PTSD. But it didn't matter: the program slid too effortlessly into military ideals, and was such a big, important-seeming, attention-getting response to the crisis, that it attained a formidable internal momentum and quickly snowballed on the basis of its own overheated promises.

Or, phrased differently: Imagine Marty Seligman and Patricia Resick competing for the same giant military contract. During Seligman's presentation, he explains how his idea, Comprehensive Soldier Fitness, will help reinforce values the Army already holds dear: self-possession, hard work, respect for and trust in authority. The trainings can slot right into soldiers' other responsibilities. Soldiers can be trained up as Master Resilience Trainers. A relatively simple, universal intervention will make the military stronger in an easy, convenient way that won't interfere with anything. It will save lives. Best of all, adopting this program will allow the Army to broadcast out to America inspiring scenes of soldiers receiving life-enhancing training, and of Master Resilience Trainers fanning out throughout the Army, imparting these messages at the unit level. And as a result, countless tragedies will be averted; this is a remarkable, revolutionary opportunity to nip Army PTSD and suicidality in the bud, en masse.

Then Resick gets up to make the case that the grant should go to her and to her cognitive processing therapy approach. This is actually less straightforward than what Marty just posited, she explains. Post-traumatic stress disorder isn't about a lack of optimism, or about a failure to "hunt the good stuff." It's much more complicated than that, and there's no evidence it can really *be* prevented, and treating it involves carefully unpacking soldiers' thought patterns and, in many cases, undoing the military's very own teachings. At the end of the day, if you send young soldiers into deadly situations and allow terrible things to happen to them, you need to approach the aftermath in a careful, responsible, evidence-based way. There's no simple solution here, no quick fix. Trauma is trauma, and it's ugly and takes time to unpack. And, if she's being honest, she can't really claim, as Marty did, that beefing up the Army's investment in cognitive processing therapy will bring with it PR opportunities. These stories *aren't* inspiring; they involve young men sitting with a therapist talking about the worst days of their lives and their lingering feelings of guilt and anger about what happened on those days. Sometimes they're crying. It's hardly ever photogenic.

Who do you think gets the contract?

# 5

# WHO HAS GRIT?

It might surprise you to find out how little evidence there is to support the idea that boosting students' "grit"—their propensity to tenaciously attack difficult problems they encounter rather than give up—is a reliably effective way to improve their school performance or to close long-standing education gaps. After all, you've probably heard otherwise. Grit is everywhere. By the time you read this, it will have been a golden child of the world of education for well over a decade. It's a sexy, appealing idea: grit predicts success, grit can be measured, and grit can be improved.

Grit's popularity is largely due to the work of the concept's inventor and chief evangelist, Angela Duckworth. A MacArthur-grant-winning social psychologist at the University of Pennsylvania, she has made very bold claims about the importance of grit for years, and those claims have been echoed by other big names, too. In her 2013 TED Talk, which has almost twenty-one million views as of August 2020, she lays out the basic theory of grit in a concise, compelling manner:

> My research team and I went to West Point Military Academy. We tried to predict which cadets would stay in military training and which would drop out. We went to the National Spelling Bee and tried to predict which children would advance farthest in competition. We studied rookie teachers working in really tough neighborhoods, asking which teachers are still going to be here in teaching by the end of the school year, and of those, who will be the most effective at improving learning outcomes for their students? We partnered with private companies, asking, which of these salespeople is going to keep their jobs? And who's going to earn the most money? In all those very different contexts, one characteristic emerged as a significant predictor of success. And it wasn't social intelligence. It wasn't good looks, physical health, and it wasn't IQ. It was grit.[1]

In her talk, Duckworth also presents grit as a new way of looking at the old problem of school achievement: "In education, the one thing we know how to measure best is IQ. But what if doing well in school and in life depends on much more than your ability to learn quickly and easily?"

She made similar claims elsewhere. "My lab has found that this measure beats the pants off I.Q., SAT scores, physical fitness and a bazillion other measures to help us know in advance which individuals will be successful in some situations," she told *The New York Times*.[2] The cover of her bestselling 2016 book, *Grit: The Power of Passion and Perseverance*, has the sort of blurb most publicists could only dream of. "Psychologists have spent decades searching for the secret of success," enthuses Daniel Gilbert, the Harvard psychologist and happiness researcher, "but Duckworth is the one who found it."[3] *The* secret of success.

There are different versions of Duckworth's grit scale, but regardless of which one you use, taking it is easy: You can measure your grit in about three minutes simply by filling out a ten-item version on Duckworth's website—items like "I finish whatever I begin," "I am diligent. I never give up," "New ideas and projects sometimes distract me from previous ones," and "I have difficulty maintaining my focus on projects that take more than a few months to complete."[4] For each one, you mark whether the statement in question is "Very much like me," "Not like me at all," or something in between, and when you're done the website spits out your grit score. I got a 2.4 out of 5.0, which is extremely low. "You scored higher than about 10% of American adults in a recent study," I was informed by the presumably unimpressed algorithm. Under the hood of these tests, two different grit "subfactors" are being measured: *perseverance*, or the extent to which someone doesn't get discouraged by challenging circumstances, and *consistency of interest* (sometimes referred to as passion), or the extent to which someone doesn't flit around from thing to thing.

In their key publications on the topic, Duckworth and her colleagues have sought to correlate individuals' grit scales with various life outcomes and to see whether grit predicted success better than other, more established measures. One important early study from 2011 found that "grittier competitors triumph at the national spelling bee."[5]

In another key paper, this one from 2014, a team that included Duckworth, Lauren Eskreis-Winkler, Elizabeth Shulman, and Scott Beal took online samples of West Point cadets, high school juniors in the Chicago Public Schools, and others, and found that grit predicted "retention in the military, the workplace, school and marriage," as the paper's title put it.[6] Duckworth described the Chicago schools result in her TED Talk as follows: "Turns out that grittier kids were significantly more likely to graduate, even when I matched them on every characteristic I could measure: things like family income, standardized achievement test scores, even how safe kids felt when they were at school." In her talk, she argued that grit is particularly important "for kids at risk for dropping out"—meaning it is relevant not just for those seeking to succeed in "super-challenging settings" (West Point, elite spelling bees) but for less privileged or accomplished people as well.

The media have helped spread the idea that Duckworth discovered something new and exciting, if the mostly favorable coverage on NPR, in the *Times*, and in a host of other big-name media outlets is any indication. Her book has been a long-term bestseller. And the excitement has filtered down into some schools themselves. Grit satisfies what the *Times* described in early 2016 as a "recent update to federal education law [that] requires states to include at least one nonacademic measure in judging school performance."[7] The Obama Department of Education expressed a lot of enthusiasm about grit,[8] and *The Sacramento Bee* reported in 2015 that some schools in California were giving students a "grit" grade. Charter schools have become quite keen on the concept.[9] For example, the Knowledge Is Power Program (KIPP), an admired network of more than two hundred charter schools around the country, has adopted grit as one of the seven foundational "character strengths" it seeks to foster in its students.[10] Many other independent schools, too, are exploring how to integrate the measurement and fostering of grit into their classrooms and curricula, and it has become a popular buzzword among education specialists.

Duckworth didn't just argue that all else being equal, effort matters; such a claim wouldn't have gotten much attention. Rather, she argued that she had developed a new, uniquely useful scale for *measuring* crucial traits and that this scale pointed toward new opportunities for improving performance in a wide variety of domains, especially the classroom. It's important to note that she doesn't appear to have ever explicitly claimed that she had discovered a reliable way of increasing grit. At one point in her TED Talk she said, "Every day, parents and teachers ask me, 'How do I build grit in kids? What do I do to teach kids a solid work ethic? How do I keep them motivated for the long run?' The honest answer is, I don't know." And yet, as we'll see, once grit became a full-blown cultural phenomenon, it became difficult to control the messaging around it. Duckworth's own claims—claims that, as we'll also see, can be reasonably described in some cases as exaggerations or oversimplifications of what her research actually found—helped the concept acquire considerable currency in the minds of millions of Americans, including some very influential ones.

GRIT'S ORIGIN STORY IS now well-known: Duckworth's interest in the subject stemmed in part from her time as a young teacher, which followed stints as the head of a nonprofit and as a consultant at McKinsey. She noticed that there wasn't always a clear correlation between students' natural intelligence and their performance in her classes. The kids who appeared to have the highest IQs often weren't the ones who got the best grades. Some sort of noncognitive force seemed to pull some kids up toward higher grades than one might expect and hold others down a little beneath their potential.

Duckworth began her formal study of grit under none other than Martin Seligman, the godfather of positive psychology. She applied to be a doctoral student at Penn in her thirties—relatively late in life for such a move—and in her application she wrote that "the problem, I think, is not only the schools but also the students themselves. Here's why: learning is hard. True, learning is fun, exhilarating and gratifying—but it is also often daunting, exhausting and sometimes discouraging . . . To help chronically low-performing but intelligent students, educators and parents must first recognize that character is at least as important as intellect."[11]

Grit, to be sure, was far from a new concept in American life. It has a long history as an American ideal, and it has often been presented as a quality conspicuously wanting in *privileged* people. As the historian of education policy Ethan W. Ris explains in an excellent short history of the subject published in the *Journal of Educational Controversy* in 2015, Americans have long feared that the more moneyed among them—particularly the young and moneyed—lack a certain gumption possessed by the striving poor.[12] Appropriately, the very first deployment of "grit" in this sense targeted that paragon of entitlement, the layabout European intellectual. "His main deficiency was a lack of grit," wrote Nathaniel Hawthorne of a British poet in 1863.

In the late nineteenth century, Ris explains, the arrival of the Gilded Age brought with it fears of "affluenza," or "the disadvantages caused by childhood advantage." To counter this, "the elite boarding schools established in New England during this time proudly advertised their Spartan quality of life, complete with cramped quarters and cold showers."[13] Popular literature took up the mantle of grit with great enthusiasm, from the "poor but gritty" Huckleberry Finn on to an endless succession of novels, most famously those of Horatio Alger Jr., which followed "a set formula: an impoverished boy, frequently abandoned or orphaned, endures a series of hardships and due to his own strong character attains a respectable middle-class position as a clerk or professional."[14]

Of course, this sort of literature contains not only an encouraging message—poor kids who work hard can claw their way into the middle class—but a darker one, too: middle-class or rich kids who *lack* grit could easily see their social station slide. The wish to inculcate grit has long been connected to a fear of slipping out of the more comfortable socioeconomic classes as a result of privilege and indolence, in the face of fierce competition from below. *Grit*, "a national weekly newspaper published continuously from 1885 to 1993 [that] focused on positive news from rural communities," was delivered, via paper routes, by middle-class children and teens to help them "gain the trait that graced its masthead."[15]

In Ris's telling, interest in grit waxed and waned and mutated throughout the twentieth century. There was a marked decline in the 1960s and 1970s, for a variety of reasons ranging from relatively low levels of inequality to the social fractures caused by the Vietnam War. It wasn't until around 2010 that grit burst back on the scene. This was largely thanks to Duckworth, though Ris believes the concept was also buoyed by a renewed national interest in poverty and inequality.

Duckworth's primary innovation was to scientize grit by developing a scale to measure it. And she soon found useful applications for that scale. In her second year as a doctoral student, she spent some time at the U.S. Military Academy at West Point, digging into a tricky question that was raised about cadets there: Why did some, but not others, successfully traverse a notoriously difficult seven-week orientation period known as Beast Barracks (or the Beast for short)? Grit, according to Duckworth, "turned out to be an astoundingly reliable predictor of who made it through and who did not," performing significantly better than the Army's own Whole Candidate Score instrument.[16]

Duckworth first administered her grit scale to West Point cadets in 2004. Before long, she'd applied it to spelling bees and high school graduation rates, and the idea had caught on in the mainstream. She gave her TED Talk in April of 2013, and five months later, at forty-three, she won her MacArthur grant for "clarifying the role that intellectual strengths and personality traits play in educational achievement."

The evidence for her strongest claims about grit's efficacy, though, still hasn't arrived. A decade and a half after the concept's introduction, it has not been established that grit is a genuinely useful concept that tells us much that we didn't already know—or that it can be boosted, anyway.

AS DUCKWORTH AND HER COLLEAGUES acknowledge in their very first paper on grit, personality psychologists already have a concept that seems similar: conscientiousness. Conscientiousness is a component of the popular "OCEAN" model of personality, which has existed in different forms since the 1960s and began to really catch on in the early 1990s. The OCEAN model claims that individual differences in personality can be usefully captured and organized through the "big five" rather self-explanatory measurable traits: openness (to experience), conscientiousness, extraversion, agreeableness, and neuroticism. This model has left a large mark on personality psychology, in part because it raises useful questions that researchers have subsequently investigated, ranging from the extent to which variation in these traits is caused by nature versus nurture—one 2015 meta-analysis estimated the answer is about 40 percent genetics, 60 percent environment[17]—to whether and to what extent various traits correlate with success in work, relationships, and other settings. Naturally, some of this research has centered on education; conscientiousness, in particular, has been found to correlate significantly with school performance, though as we'll see it explains only a fairly small part of why some students are higher achievers than others.

Duckworth seemed to realize early that grit bore certain similarities with—and was in a sense "competing" with—conscientiousness. If grit turned out to predict school performance, but only a third as much as conscientiousness (for example), it might be seen as unimportant. She and her colleagues theorized in their first 2007 paper that their new construct was measuring something a bit different:

"Grit overlaps with achievement aspects of conscientiousness but differs in its emphasis on long-term stamina rather than short-term intensity."[18] And indeed, some of the items on the grit scale, like "I have difficulty maintaining my focus on projects that take more than a few months to complete," are clearly geared at capturing an element of long-term single-mindedness that, in the view of Duckworth and her colleagues, conscientiousness doesn't.

Discussions of the potential overlap between grit and conscientiousness have mostly played out in the scientific literature rather than in pop-science treatments of grit: "conscientiousness" doesn't even come up in Duckworth's TED Talk.[19] There and in other public-facing presentations of the idea, Duckworth's most eye-catching claim was that people were overestimating the importance of what many view as natural talent, as measured by things like the SAT or IQ tests, relative to hard work and stick-to-itiveness, and that grit offered a way of revealing this truth. This was the argument she was advancing when she made the claim about grit "beat[ing] the pants" off other measures.

In some cases, it seems Duckworth has oversimplified her initial grit research, or given a possibly misleading account of the results. Take her very first paper on grit, co-authored with Christopher Peterson, Michael D. Matthews, and Dennis R. Kelly. It consists of six studies seeking to correlate grit and other outcomes with success in domains like school, the military, and a national spelling bee. In some cases, there are strong-sounding claims that, in their full context, become less impressive. One of Duckworth's most famous findings, for example, centers on grit's accuracy in predicting which freshman cadets would successfully traverse Beast Barracks. Most notably, Duckworth concluded that grit did a better job predicting retention than the Army's Whole Candidate Score, a proprietary rating program drawn from a given student's academic, leadership, and physical abilities.

This finding comes from studies 4 and 5 in that original, seminal grit paper. In study 4, grit indeed outperformed the Whole Candidate

Score and a psychological instrument known as the Brief Self-Control Scale in predicting who would make it through Beast Barracks. In fact, when these and other variables were taken into account, "grit predicted completion of the rigorous summer training program better than any other predictor." The researchers got a similar result in study 5, which was structured in the same way except that instead of measuring cadets' self-control, the researchers measured their Big 5 conscientiousness. Again grit performed better than the other variables in predicting retention. And in 2009, Duckworth and Patrick D. Quinn published a replication of this result based on a different sample of West Point recruits, using a newly developed, shortened version of the grit scale.[20]

Surely, this is a victory for grit. But it's actually a very narrow one. The problem is that because the vast majority of cadets pass Beast Barracks—about 95 percent on average—the results need to be interpreted in light of a well-known statistical issue called "range restriction" having to do with samples clustered in one part of a distribution rather than spread throughout it (a random sample of NBA players will likely have range-restriction issues when it comes to height, for example, that a random sample of Americans will not). Almost everyone in the sample scored well on this outcome: they passed, rather than didn't pass, Beast Barracks. So while grit does offer some predictive usefulness, in context it isn't all that impressive: About 95 percent of all candidates made it through Beast Barracks, and about 98 percent of the grittiest ones did. Other areas of grit research involve other forms of range restriction, as Duckworth, to her credit, has acknowledged: Some of her work involves Penn students whose SAT scores are clumped near the top of the distribution, for example. And range restriction issues make it even harder than usual to generalize from these results to other settings. (In her book, Duckworth describes Beast Barracks as intensely difficult, noting that "one in five cadets [at West Point] will drop out before graduation" and that "a substantial fraction of dropouts leave" during Beast

Barracks. This is all technically true, though one could quibble with one out of twenty being "a substantial fraction," but it elides the fact that the vast majority of cadets defeat the Beast.)

Duckworth and her colleagues didn't just study summer retention in study 4; they also looked at candidates' grades, as measured by their GPAs and so-called military performance scores, or MPSs, which combine grades and other performance metrics. In these somewhat less range-restricted settings, suddenly grit was not particularly useful. It was about as useful as the self-control scale in predicting GPA and MPS but was trounced by the Whole Candidate Score. "When Whole Candidate Score and self-control were held constant," grit accounted for, at best, just 0.8 percent of the variance in MPS and none of the variance in GPA. This suggests that grit is not a particularly useful instrument for measuring academic achievement at West Point.

Duckworth's claims in other areas also merit scrutiny. According to Duckworth, grit was more useful than other measures in examining teacher performance, particularly among new ones working in challenging situations. And yet neither of the two studies Duckworth and her colleagues have published on grit and teachers fully evaluates the claim that grit outperforms other, traditional measures in this domain. In one, published in 2009 in *The Journal of Positive Psychology*, with Martin Seligman as one of her co-authors, Duckworth and her colleagues correlated "optimistic explanatory style, grit, and life satisfaction prior to the school year" with teacher performance at the end of it (as measured by students' grades—let's set aside the question of whether this is fair).[21] As it turned out, when all three variables were taken into account, grit and life satisfaction were significant predictors of performance and explanatory style wasn't. But the researchers simply didn't evaluate those traditional measures with which grit is supposedly competing, like intelligence and conscientiousness. As a result, this study cannot tell us whether grit did a better job predicting teacher performance in this cohort than these other available measures. The other study Duckworth cites to support this claim used an entirely different scale that, on its face, has very little to do with grit, and therefore doesn't tell us much about the utility of grit in this context, either (I'll relegate these in-the-weeds details to an endnote).[22]

What about those Chicago high school kids? "Turns out that grittier kids were significantly more likely to graduate, even when I matched them on every characteristic I could measure," Duckworth said in her TED Talk. True, but if you look at that study, you'll find that when all the relevant variables were adjusted for, grit accounted for just an additional 0.50 percent of the explained variance in student retention. Standardized test scores, by way of contrast, accounted for 4 percent, dwarfing the predictive utility of grit. And the researchers relied on a data set that included responses to two different instruments related to "academic conscientiousness and school motivation," but traditional, Big-Five conscientiousness wasn't measured. Finally, while the Beast Barracks finding does appear to be one of the few instances in which grit truly offers something other measures can't (range restriction issues aside), in another, similar setting grit was far less impressive. In a 2014 study, Duckworth and her colleagues tried to correlate grit and other variables with successful completion of a U.S. Army Special Operations Forces selection course. This is a more difficult task than Beast Barracks; in this particular sample, 42 percent of candidates failed to complete the course, meaning it's a situation in which the Army would likely be even keener on developing better predictors of success. When all the relevant variables were included in a statistical model, grit accounted for just about 1.8 percent of the variance in retention, compared with about 2.7 percent for general intelligence and 7.2 percent for physical fitness. In this study, again, the researchers didn't test conscientiousness, so it's hard to know whether grit would have done a better job predicting retention. Either way, this was yet another example of grit not actually outperforming those other, more "traditional" variables, even if it did offer *some* statistical usefulness.[23]

It should be clear from these results that there was never much in the literature to support either of the two ideas that launched grit on its way: that it was more useful than conscientiousness and that it seriously outperformed "traditional" measures of cognitive or, in the context of military training, physical performance. It is difficult to justify Duckworth's statement that grit "beats the pants" off older, more established measures. Many of the examples she gives consisted of studies in which the predictive usefulness of grit wasn't compared with its most obvious competitor, conscientiousness, in which grit simply didn't perform as well as traditional measures, or both.

Which leaves the concept where, exactly? The most comprehensive answer came in the form of a 2017 meta-analysis published by Marcus Crede, Michael Tynan, and Peter Harms. Crede is a reform-minded psychologist who has a keen sense of how statistics can be misused to prop up half-baked ideas. He's made it his mission to critique what he views as questionable findings in his field and has a particularly keen interest in education and workplace performance. (He also, you may recall, played a role in puncturing some holes in the most overheated claims about power posing.) In their paper, titled "Much Ado About Grit," Crede and his colleagues argue the concept suffers from some weaknesses that have gone mostly ignored but also show how, at the end of the day, there could be some utility to it.[24]

Crede and his colleagues write that there is strong statistical evidence that "the grit-conscientiousness relation may be much stronger than is commonly assumed."[25] How did prior research miss this? It is generally difficult to accurately measure something like grit or conscientiousness; these scales are fundamentally pretty noisy. This statistical noise can make their level of correlation appear lower than it actually is. But if you can correct for that noisiness, you can get a truer measure of their correlation. And "numerous [researchers] have reported correlations between grit and conscientiousness that approach unity when correcting the observed correlations for unreliability," Crede and his colleagues write.[26] In this context, "unity" simply means a situation in which the two are perfectly correlated. (The most intuitive nonstatistical way I could come up with to explain this type of statistical correction to myself was this: Imagine you have two tape measures, each warped and skewed in different ways. Each one will give a different result for your height. If you had some way of partially correcting each tape measure's level of warpedness, suddenly they would offer much more similar results.)

Both grit and conscientiousness seem to be measuring the same underlying concept, argue Crede and his co-authors. Therefore, they suggest, grit's popularity might be the result of the *jangle fallacy* in which people believe that two things that are actually the same are different simply because they have different names. That is, if Duckworth had published research showing that conscientiousness can, to a certain extent, predict academic success, other researchers would have rolled their eyes and said, "Of course, we already knew that." But by presenting a seemingly new concept with a catchy name, Duckworth might have gotten a great deal of mileage out of an idea that had been part of the literature all along (which is not to suggest that this was some sort of intentional obfuscation on her part). NPR reported in 2016 that Duckworth, responding to this critique, said "she would prefer to think of grit as 'a member of the conscientiousness family,' but one with independent predictive powers."[27]

Of course, one might say, well, maybe grit *is* just conscientiousness, but if conscientiousness itself matters to school achievement and talk of grit gets people re-excited about conscientiousness and its connection to academic success, would that be so bad? Here's where things get complicated and where it's important to separate out two different questions: whether conscientiousness predicts school achievement, and whether it can be reliably boosted.

On the first question, both conscientiousness and grit *do* appear to be correlated with school performance—somewhat. One meta-analysis, published by Arthur Poropat, found that conscientiousness and school performance were correlated at $r = 0.19$.[28] If you square that so-called *r* value and multiply by 100, that's the percentage of the variance in academic performance accounted for by grit, holding everything else constant: In this case, conscientiousness accounts for about 3.6 percent, or $0.19 \times 0.19$, of the variance in school performance. Crede and his colleagues' meta-analysis generated a similar estimate: grit was correlated with school performance at $r = 0.18$.

Is 3.6 percent or so a high number in this context? In a certain, raw sense, yes: social scientists generally consider it impressive when a briefly administered instrument can bring (even) this much predictive power to a complicated outcome like school performance. But it appears that intelligence is a much, much more powerful predictor: researchers believe that IQ accounts for about 25 percent of the variance in school grades.[29] In some studies it is higher, but even taking this potentially conservative estimate, IQ explains about seven times more of the variance in academic achievement than grit or conscientiousness. So, at least as far as school performance is concerned, the most accurate response to Duckworth's TED Talk question, "What if doing well in school and in life depends on much more than your ability to learn quickly and easily?" may be something like "Sure, intelligence isn't the whole thing, but it's a lot more important than grit or conscientiousness, as far as anyone can tell." (It's important to note that the connection between intelligence as measured by IQ and school performance varies widely by subject. One study found the "variance accounted for [by intelligence] ranged from 58.6% in Mathematics and 48% in English to 18.1% in Art and Design."[30] It shouldn't surprise anyone that some brilliant artists, for example, don't score off the charts on traditional IQ tests, because of course these tests capture only a fraction of people's potential to accomplish worthwhile things.)

As for the question of grit's malleability, there isn't much evidence of reliable, scalable interventions for increasing conscientiousness or grit. That isn't to say conscientiousness remains immutable across the life span. "Happily, many studies show that conscientiousness does change with age," Brent Roberts, a leading personality psychologist and director of the Center for Social and Behavioral Science at the University of Illinois at Urbana-Champaign, told me in an email. "And, not only does it change, but typically for the better—it goes up . . . Of course, changing slowly, incrementally, through life experiences is nice, but may provide little solace to the parent of a teenager who remains 'unmotivated.'" (Sure enough, one of Duckworth's key early papers includes a chart showing average grit differences by age that exhibits this general pattern.[31])

Is there *any* research that might provide succor to parents of gritless teens? In one personality-change meta-analysis co-authored by Roberts, he and his team found that while neuroticism could be reduced significantly and durably via therapy and similar interventions, there was little evidence that conscientiousness was similarly mutable.[32] People have wanted to instill grittiness forever, of course, but "studies focused on changing conscientiousness [specifically] are just beginning to emerge," he said in an email he sent me in November 2018. One, out of Australia, involved a group of psychologists who implemented a "personality coaching" intervention: after the study participants each chose a trait they wanted to improve, their psychologists (or "coaches") worked on it with them for a period of months. Follow-up appeared to show that both conscientiousness and the other Big 5 traits could be altered a bit.[33] But that study, which was small, entailed ten weeks of individualized coaching, which might not be scalable to an educational setting. Moreover, it was conducted not on a random sample of schoolkids but on a group who specifically chose conscientiousness as the trait they wanted improved and who therefore might have been more motivated than, for example, a teen subjected to a supposedly grit-enhancing exercise at school.

When I emailed Duckworth herself about the apparent lack of evidence that grit or conscientiousness is easily malleable, she said she didn't think such traits could be changed "overnight" but was more hopeful about longer-run efforts, and she suggested I speak to Roberts for more information. But neither Duckworth, Roberts, nor anyone else I contacted has been able to point me to a single study that demonstrates the sort of result that would instill confidence that grit is a sufficiently malleable attribute in educational settings to have earned so much attention in recent years, especially in light of the fact that its correlation with school performance is not particularly impressive in the first place.

Now, all this said, Crede and his colleagues did find some worth in grit. Their paper suggests that one of grit's two facets, perseverance, offers reasonably meaningful correlations with the various outcomes Duckworth and others have studied. The other, consistency of interest (covered by statements like "My interests change from year to year"), isn't as useful. Therefore, they argue, the "primary utility of the grit construct may lie in the perseverance facet." This is a useful point that others have made more intuitively: Does it really make sense to say that people aren't gritty if they sometimes flit around from thing to thing? Many successful serial entrepreneurs, for example, are capable of wisely abandoning a sinking project before wasting too much more time on it, but also of devoting a hundred hours a week to projects that appear to be bearing fruit. And plenty of other vocations, too, seem to require this sort of flexibility; as Daniel Engber wrote about the grit controversy in *Slate*, "As a journalist, I thrive on flexibility, flitting around from one topic to another; I don't believe my job lends itself to grit."[34] (For what it's worth, I feel similarly.)

When Crede and his colleagues ran the numbers, they found that while grit did not predict academic performance once conscientiousness was taken into account, the perseverance of effort subfactor actually did provide an incremental boost. This means that part of the grit scale could, in fact, offer a slight improvement over traditional conscientiousness in predicting academic performance. In 2018, another research team consisting of David Disabato, Fallon Goodman, and Todd Kashdan, published a *Journal of Personality* study of a large online sample of thousands of people from around the world, finding that while "perseverance of effort was moderately to strongly related to subjective well-being, beliefs about well-being, and personality strengths . . . consistency of interests had weak or negative correlations with these outcomes."[35] This stark divide between the grit scale's two subscores constitutes an interesting twist in the story and could open up some new research avenues. In light of all the other problems with grit, however, it doesn't really change the overall situation.

Finally, in July 2020, as I was finishing up this book, an important study was published that sought to fill in some of the gaps in the grit literature. Chen Zissman and Yoav Ganzach, of Tel Aviv University and Ariel University in Israel, published an article in the journal *Social Psychological and Personality Science* in which they examined a large, representative sample of Americans and found that "intelligence contributes 48–90 times more than grit to educational success and 13 times more to job-market success."[36] Conscientiousness, meanwhile, was twice as useful at predicting success as grit was. This study, Ganzach told me in an email, was the first to analyze grit among a representative sample (meaning problems like range restriction are much less likely to come up). All told, it bolsters the case that while grit might be useful in certain very specific domains, it is not, broadly speaking, a particularly helpful concept for predicting who will succeed and who will not—or at least it doesn't beat the tools we already have at our disposal.

TO HER CREDIT, DUCKWORTH has been significantly more candid and transparent than other researchers who have found their ideas under scrutiny, and she has been generally open about the limitations of the research. Even as Duckworth was promoting her book, for example, she started speaking out about some of the questionable ways the concept was being applied. In a *New York Times* column published in March 2016, she cautioned against grading students on their level of grit, as several California schools were doing, in part because the research on the subject was too young. "We're nowhere near ready—and perhaps never will be—to use feedback on character as a metric for judging the effectiveness of teachers and schools," she cautioned in her op-ed. "We shouldn't be rewarding or punishing schools for how students perform on these measures."[37] Duckworth has expressed frustration at the fact that she had, to a certain extent, lost control of the grit narrative. As my then colleague Melissa Dahl wrote in 2016, after interviewing her, "It's a strange thing, Duckworth said, to have played a significant part in the creation of an idea, only to have that idea run away from you and create a life of its own."[38] Again, in both her book and her TED Talk, she explicitly said she was not aware of specific interventions that could boost grit but that she hoped some could be developed.

But Duckworth still defends her early claims for the usefulness of the grit scale in ways that feel like a bit of a stretch. When I emailed her to ask about her quotation in *The New York Times* that "my lab has found that this measure beats the pants off I.Q., SAT scores, physical fitness and a bazillion other measures to help us know in advance which individuals will be successful in some situations," she stood by it, saying that the context of those remarks was her West Point and teacher research. But as we've seen, there are serious questions about the utility of grit in these two contexts, relative to other, more established measures. It is unlikely that the average *Times* reader or TED Talk viewer would come away from Duckworth's claims about grit understanding just how crimped and conflicted some of the research on the topic is.

Whatever one thinks of Duckworth's handling of this controversy, there is ample reason to regard the larger vogue for grit with skepticism. Grit appears to be quite similar to conscientiousness;

there's little available evidence conscientiousness can be durably boosted in school settings; and the correlation between conscientiousness and academic performance isn't that strong, anyway, in most contexts. So an obvious question pops up: Is *this* the thing for schools to focus on? It's not as if school administrators are lacking for other options. As Crede and his colleagues point out in their meta-analysis, "Study skills and study habits, adjustment to college, and class attendance are . . . far more strongly related to academic performance and retention than grit, and there is sound evidence that interventions can improve students' standing on these constructs (especially for study skills and habits)"—evidence we lack for grit.[39] That stuff is just a lot less exciting. If you give a talk titled "Study Skills Are Important and Improvable," there will be some empty seats.

Novelty, then, can partly explain why America has fallen so thoroughly for a concept with so little hard evidence behind it. But to truly understand the appeal of the grit story line requires understanding two aspects of our national educational conversation: the view that American schools should imbue character in their pupils, and the (related) idea that schools are crucial sites for ameliorating or solving problems caused by forces beyond their walls.

AMERICANS AREN'T UNIQUE in their habit of linking education with character. For as long as human beings have talked about education, intellectual and moral leaders have insisted that it isn't enough for schools to impart knowledge and problem-solving skills to students. No, schools should also teach *character*. That is, they should teach students how to be virtuous, kind, faithful citizens. This idea can be traced back millennia. "Who then shall I call educated?" asked the ancient Greek orator Isocrates, considered one of the fathers of education, in 342 B.C. "Those who manage well the daily circumstances of their lives, who possess accurate judgment and who rarely miss the proper course of action; those who are decent and honorable, good natured, slow to take offense, disciplined in their pleasures; brave under misfortune and unspoiled by success. Those who have a character which is in accord not with one of these qualities, but with all of them—these are the wise and complete men."[40] Unless you want an Athens crawling with unwise, incomplete men, in other words, you better mix some moral instruction in with all that math.

It's unsurprising, then, that when American education reformers, led by the pioneering Horace Mann, began to lay the groundwork of the modern school system in the nineteenth century, character was at the forefront of their effort. While in later years education would come to be tightly linked with prosperity, Mann and his contemporaries were "reluctant to portray education as a way to promote worldly gain," as the education researcher David Labaree writes.[41] Mann had something bigger in mind as he sought to streamline a hodgepodge of public and private schools into a unified "common school" system; he wanted to forge *citizens*. "It may be an easy thing to make a republic," he wrote, "but it is a very laborious thing to make republicans; and woe to the republic that rests upon no better foundations than ignorance, selfishness, and passion!"[42] From the first days of what would become the modern American school system, character development was seen as a paramount goal.

Initially, a great deal of American character education came straight from the Bible; who in nineteenth-century America could argue with that? Soon some of the focus shifted to so-called McGuffey Readers, named after the educator William Holmes McGuffey. These readers, which sold around 120 million copies between 1836 and 1960—on the same level as Webster's dictionary and the Bible—were geared at instilling a sense of duty and patriotism. "No textbooks since have equaled these readers," the Wisconsin state senator John E. Cashman told the Associated Press in 1935, arguing that a revived interest in them could help snap the nation out of the Depression. "Every lesson had a moral to it. They taught love of country and didn't criticize the government. There's too much criticism and not enough patriotism given the children now."[43]

The question of character education got more complicated as the twentieth century wore on and the United States exploded in size, grew increasingly diverse after multiple waves of immigration, and eventually established itself as the preeminent world power. In the 1960s, character education receded from the scene a bit—as one account put it, "the rapidly intensifying pluralism of American society and the increasing secularization of the public arena forced many schools away from their once central role as moral and character educators."[44] (You'll recall that according to Ris, there was a similar lull, for similar reasons, in societal interest in grit during this period.)

But not for good. One major recent wave of interest in character education began with Ronald Reagan's moral crusader of an education secretary, William Bennett, who wanted schools to foster moral character in their students more aggressively and even wrote a book, *The Book of Virtues*, on how to help nudge children down that path. Bennett had a very particular, quite conservative view of what this meant, of course—he once said that the difference between right and wrong cannot be taught "without reference to religion"[45]—but the general idea of using schools to instill character was resuscitated as a bipartisan political talking point as a result of his advocacy and President Reagan's embrace of the concept. Two later presidents, Bill Clinton and George W. Bush, each promised to triple the funds allotted to character-based education programs—a sure sign character education had become a full-blown fad.[46] "I challenge all our schools to teach character education, to teach good values and good citizenship," said Clinton in his 1996 State of the Union address,[47] and he reprised that theme in his 1997 address: "Character education must be taught in our schools. We must teach our children to be good citizens."[48]

Elite interest in character education produced a robust market for supposedly character-enhancing educational interventions. And because character means so many different things, sometimes the breadth of the mission led to alphabet soup: One program, called AEGIS, or Acquiring Ethical Guidelines for Individual Self Governance, focused on the six core areas of "worth and dignity, rights and responsibilities, fairness and justice, effort and excellence, care and consideration, and personal integrity and social responsibility." AEGIS sought to inculcate those core precepts through its SMILE method—that is, Stimulating interest, Modeling the concept in question, Integrating the concept, Learning more about the concept via homework assignments, and, finally, Extending it to real life.[49]

The Character Education Curriculum, on the other hand, sought to teach not six but *twelve* universal values to children: honor, courage, convictions, honesty, truthfulness (meaning honesty?), generosity, kindness, helpfulness, justice, respect, freedom, and equality. While that program lacked any "elaborated teaching or learning model," its materials did helpfully explain, as James S. Leming summed it up in 1997, "that character consists of three equally important components: *knowledge*—awareness, self-knowledge, and decision-making; *feeling*—conscience, respect, caring, and empathy; and *action*—good habits, behavior, and determination."[50] (For those keeping track at home, that's twelve universal values, which rest on three components, which in turn consist of ten total subcomponents.)

These were not, generally speaking, scientifically informed programs. Leming notes that "few if any of the curricula presented [here] offer what can be called a fully elaborated statement containing the underlying theoretical rationale for the curriculum."[51] And as these programs' popularity and reach expanded in public schools, they were drawn into culture war brawling as well. During the Clinton years, writes the journalist Paul Tough, "the right suspected that character-education initiatives were a cloak for creeping political correctness, and the left suspected the initiatives were hidden attempts at Christian indoctrination."[52] Again, while everyone agrees kids should have character, the question of *which type* of character seems to consistently invite cacophony.

On those rare occasions when character education programs have been evaluated rigorously, the results have been unimpressive. In 2010, the Department of Education published a report that found just about nothing to recommend a handful of such programs designed for elementary school students. "On average, the seven programs did not improve students' social and emotional competence, behavior, academic achievement, and student and teacher perceptions of school climate," wrote the authors. "In addition, although the numbers of schools and students in each program were not always sufficient to support firm conclusions at the program level, the patterns of estimated impacts for each program were largely similar: students' outcomes were not affected."[53] We also saw earlier that the Strath Haven Positive Psychology Curriculum, which benefited from almost $3 million worth of government funding for evaluation purposes, didn't appear to improve students' "character strengths," as Marty Seligman and his colleagues called them, either.

While character educators will sometimes point to some program or another and claim it holds promise, the general consensus is that no one has come up with a good, empirically validated way to accomplish the somewhat hazy goal of giving kids more "character." That's true whether you measure these programs' success by academic outcomes or by, to borrow the DoE report's language, "social and emotional competence"—which is really a fancy way of referring to one set of "character" concerns. If these programs really *did* make kids more kind or honest (for example), then they might well be worthwhile, even if they didn't improve academic achievement, but they simply don't appear to do so.

THE CHARACTER EDUCATION CRAZE points to a larger reality about the American politics of education: we put an unreasonable burden on our school systems, expecting them to solve problems that may be beyond their means. Some of this is due to the particular ways America's educational and social welfare structures evolved. "In the late nineteenth and early twentieth centuries, while other nations were introducing unemployment, old age, and health insurance, the United States was building high schools for a huge surge in enrollment," wrote the historian and social theorist Michael B. Katz in a 2010 article in the democratic socialist journal *Dissent*. Historically, he notes, "European nations were about a generation behind the United States in expanding secondary education; the United States was about a generation behind Europe in instituting its welfare state."[54] In the United States, then, schools are expected to do more social welfare work than they are elsewhere, which could help further fuel the notion that they should be imbuing not just academic skills and knowledge but character, too.

But many people from all over the political spectrum have long argued that it isn't realistic to expect schools to fulfill one of the key purposes of welfare systems: closing inequality gaps. By the time a child enters kindergarten, after all, he or she has already gone through an absolutely crucial half decade of development. To ask how a five-year-old is likely to do later in life is to be faced with the depressingly unfair fact that the answer has to do with factors well beyond that five-year-old's control—factors that have likely already thumbed the scale quite a bit. At five, a child either has or hasn't had access to sufficient nutrition, to sufficient cognitive stimulation, to sufficient family stability, and to protection from the sorts of adverse childhood events that have now been linked, in later life, to everything from smoking to heart disease.

In a 2016 article in *The Atlantic* adapted from his book *Helping Children Succeed: What Works and Why*, Tough explains that "neuroscientists have demonstrated with increasing clarity how severe and chronic stress in childhood—what doctors sometimes call toxic stress—leads to physiological and neurological adaptations in children that affect the way their minds and bodies develop and, significantly, the way they function in school." Toxic stress can also "disrupt the development of what are known as executive functions: higher-order mental abilities that some researchers compare to a team of air-traffic controllers overseeing the workings of the brain," such as "working memory, attentional control, and cognitive flexibility."[55] This could help explain why preschool programs work: by providing nurturing, stable environments, they reduce children's stress levels, fostering the development of executive functions. (There is a tremendous amount of debate over the effectiveness of such programs; it is arguably one of the most complicated and contentious contemporary social science debates. But while certain high-profile preschool efforts have failed to generate positive results, overall, the data point in a positive direction.[56])

Many, many children do not have access to such a program, or to parents who are in a position to provide a stable, nurturing environment in lieu of one. If significant harm has been done to a child by the time he or she enters kindergarten, schools usually can do little to reverse it. They can do even less to narrow the increasing gap between haves (who grow up with stable homes, sufficient nutrition, and nurturing and highly educated parents and who arrive at school ready to take advantage of educational opportunities) and have-nots (who grow up with unstable homes, food insecurity, and overworked or absent or abusive parents). That would be true even if haves and have-nots attended the same schools, but they don't, of course: compounding the unfairness of the situation, kids living in rich areas are likely to have access to much better schools, whether the metric is teacher quality, physical plant quality, or the likelihood of in-school violence.

If you look at all this unblinkingly, it seems strange to imagine schools, as they exist in the United States, could ameliorate much of this damage, especially because they're tasked not "just" with closing the inequality gap but also with imparting all sorts of academic skills and knowledge. That the United States performs significantly worse than other wealthy, developed countries on key metrics pertaining to infant and childhood poverty, health, and inequality in general suggests a particularly challenging situation.[57]

It shouldn't come as a surprise that even education experts who are philosophically amenable to government-funded welfare efforts often doubt that American education can in itself enable social mobility. In his compelling book *Someone Has to Fail: The Zero-Sum Game of Public Schooling*, for example, the Stanford education professor David Labaree goes so far as to argue that under the current system one should *never* expect schooling to close inequality gaps, because relatively well-off parents can always get their kids the newest credential, while less well-off ones, no matter how hard they work or how noble their intentions, cannot. "School reform can only have a chance to equalize social differences if it can reduce the educational gap between middle-class students and working-class students," he writes. But because "we aren't willing to cap the educational opportunities for the advantaged, . . . increasing those opportunities for the disadvantaged will make no difference. As long as both groups gain more education in parallel, then the advantages of the one over the other will not decline. And that is exactly the situation in the American school system."[58]

All of this offers a strong reason to be skeptical of the claim that grit instruction—or any sort of similar effort, really—could make much of a dent in the massive problem that is American educational inequality. But I'd go a step further: It may be unfair to poor kids to focus on grit. Doing so reflects a blinkered understanding of how inequality operates and perpetuates itself.

MUCH OF DUCKWORTH'S BOOK consists of vignettes from "grit exemplars" she has interviewed over the years, from the hall-of-fame quarterback Steve Young down to promising young academics. The idea seems to be that their stories can tell us something important about grit, and therefore about success. But we don't really hear anything about hardworking, gritty, resilient people who don't get as far as they would like to, or who fail spectacularly; the losers are nowhere to be seen. The book suffers from what social scientists call survivor bias: Duckworth draws all her anecdotal lessons from a pool of exemplars chosen *because they are winners*, meaning a considerable portion of the greater picture is obscured. So when one grit exemplar explains that at difficult points in his professional life, "I'd say to myself, 'Just keep working hard and learning, and it will all work out,'" it's hard not to think of the many, many people who have had these same thoughts but ended up stuck in middle management, laid off, or worse—not as the CEO of Vanguard.[59]

As its title suggests, another book, *When Grit Isn't Enough: A High School Principal Examines How Poverty and Inequality Thwart the College-for-All Promise*, takes a very different approach. Written by Linda F. Nathan, who has decades of experience in the Boston school system and is the headmaster of the Boston Arts Academy, *When Grit Isn't Enough* focuses on the one-third of students from her school who don't end up graduating from college, even though an impressive 94 percent of them are accepted to one.[60] It contains many agonizing examples of students who exhibit tremendous amounts of what can only be called grit, only to collide headlong with the reality that unless you come from a family with certain resources—in terms of both social capital and regular old capital-capital—there are many land mines you must avoid as you inch your way toward a bachelor's degree.

Take Shanita, who "despite being the valedictorian of her class . . . lost a scholarship because she didn't send in her deposit to hold her place."[61] Or Carissa, who got into her out-of-state dream school but had to drop out after a year and a half because her mother didn't understand she had to fill out financial aid forms every year, and Carissa wasn't able to get a hold of the tax documents she needed in order to do it herself. In the middle of her sophomore year, she got an email informing her that she simply couldn't come to class anymore. And that was that. She ended up moving back home to Boston, working, and taking courses part-time at Northeastern University—a significantly more difficult path.[62] These and the book's many other stories show just how easily some highly talented teens from underprivileged backgrounds can get knocked off course by events wealthier kids take in stride—and that when this does happen, it isn't due to a lack of grit.

Nathan argues that these stories reveal that "it's not that grit is unimportant; it's just not sufficient."[63] She sums things up as follows:

> In middle- and upper-middle-class families, an invisible safety net typically surrounds young people planning to go on to college. There is usually a family member or friend who will step in and remind a student about the intricacies of student loans and deadlines, or the many requirements for staying registered once enrolled, or issues that can arise with housing. These kinds of conversations are commonplace at many dinner tables and part of numerous e-mail correspondences. However, if you are a lower-income student and you miss one or two e-mails or have a change in your advisor, you may find your dreams derailed. It may be tempting to dismiss the examples above as ineptitude or carelessness on the part of individual students, but why must there be different rules, expectations, and outcomes for low-income versus middle- or upper-income students?[64]

My own background makes me sympathetic to Nathan's skepticism of contemporary grit talk. I was the very model of a gritless adolescent (Duckworth's test was accurate, in my case). I phoned in homework that I didn't find intrinsically interesting, I couldn't fill out a complicated, boring form on time to save my life, and I had little idea how to fully survive, in the real world, without some degree of parental assistance, until well into my twenties. And none of these shortcomings mattered at all. I was enmeshed in a family and an elite public school system that would all but drag me to the gates of college and drop me off there when it was time, no matter what. After academic hiccups that could have proven quite costly to Nathan's students, who were just twelve miles or so away, I inevitably got second and third and fourth chances. Thanks to my school's grade-curving policies, when I got a C in an honors course due to laziness, it was worth the same, GPA-wise, as one of my peers' hard-earned Bs in non-honor classes (if memory serves).

My tale is far from unique. Ask around in *any* wealthy suburb and you'll find similar stories: parents filling out college applications for their kids, or hiring someone else to; endless hours spent in test prep and private tutoring; tuition not much of a concern because of the accumulated wealth. Some kids in wealthy suburbs do develop grit nonetheless, and some don't. Perhaps that partially accounts for who gets into Harvard and who "just" gets into a competitive but non-elite school (although latent ability surely plays a large role there as well). But the point is that at a fundamental level grit doesn't really matter in such a context, because, as Nathan points out, strong safety nets protect wealthier kids from falling off the college track permanently, and college is one of the most important prerequisites for comfortably supporting oneself in the twenty-first-century American economy (albeit one of an increasingly "necessary but not sufficient" variety).

What's interesting is that some of the stories in Duckworth's own book clearly reflect the limitations of a grit-centric approach, even if she doesn't quite say so explicitly. For example, toward the end she writes, "Not every grit paragon has had the benefit of a wise father and mother, but every one I've interviewed could point to *someone* in their life who, at the right time and in the right way, encouraged them to aim high and provided badly needed confidence and support." Then she introduces Cody, one such exemplar. He was born in a prison and quickly handed over to a less-than-nurturing grandmother but eventually made it all the way to MIT, in no small part because of "Chantel Smith, an exceptionally wise math teacher who all but adopted him":

> It was Chantel who paid for Cody's driving lessons. It was Chantel who collected a "college dorm fund" to pay for the supplies he'd need once he moved. It was Chantel who mailed sweaters, hats, gloves, and warm socks to him for the cold Boston winters, who worried about him every day, who welcomed him home each holiday break, who stood by Cody at his grandmother's funeral. It was in Chantel's home that Cody first experienced waking on Christmas morning to presents with his name on them, where he decorated Easter eggs for the first time, and where, at the age of twenty-four, he had his first family birthday party.[65]

It sounds as if the financial and emotional and logistical support Chantel provided to Cody were immensely helpful and comparable to what kids from more stable backgrounds get from one or both parents. (I didn't have to ask for any of what Chantel gave Cody, and it would have been viewed by others in our community as shocking if I lacked any of it.) It's also unclear what this has to do with grit—the concept this story is supposed to illustrate. If Cody had grit but lacked Chantel (and her considerable financial resources), where would he have ended up?

A similar message can be taken from the end of another one of Paul Tough's books. He concludes *How Children Succeed: Grit, Curiosity, and the Hidden Power of Character* with some recommendations that could just as easily be proffered by anyone with a deep concern about poor children, regardless of their interest in grit, or conscientiousness, or noncognitive skills, or whatever you want to call these "character" concerns:

> We could design an entirely different system for children who are dealing with deep and pervasive adversity at home.

> It might start at a comprehensive pediatric wellness center . . . It might continue with parenting interventions that increase the chance of secure attachment, like Attachment and Biobehavioral Catch-up, or ABC . . . In prekindergarten, it might involve a program like Tools of the Mind that promotes executive-function skills and self-regulation in young children. We'd want to make sure these students were in good schools, of course, not ones that track them into remedial classes but ones that challenge them to do high-level work. And whatever academic help they were getting in the classroom would need to be supplemented by social and psychological and character-building interventions outside the classroom . . . In high school, these students would benefit from some combination of what both One-Goal and KIPP Through College provide—a program that directs them toward higher education and tries to prepare them for college not only academically but also emotionally and psychologically.
>
> A coordinated system like that, targeted at the 10 to 15 percent of students at the highest risk of failure, would be expensive, there's no doubt. But it would almost certainly be cheaper than the ad hoc system we have in place now. It would save not only lives but money, and not just in the long run, but right away.[66]

What all these programs are doing is, again, providing poor kids with the stuff most rich kids already get. Maybe there aren't any shortcuts here.

So it could be that grit hype caught on because of its seductive promise to spare us a great deal of trouble. A serious effort to make life less unfair for neglected kids would likely require enacting bigger, more ambitious redistributive social programs—social programs that are very unlikely to be enacted given the state of twenty-first-century American politics. Grit, by contrast, is a quick fix.

OR MAYBE, AGAIN, at the marketplace-of-ideas level, grit was never *really* "for" poor people anyway. That's the argument of Ethan Ris, whose useful history of the concept I quoted at the beginning of this chapter. "At first glance, the latter-day importance of grit seems to be entirely about improving the prospects of students in disadvantaged communities," he writes. But "the more probable subjects of today's grit instruction are . . . much wealthier ones."[67] Duckworth has shown interest in kids from low-income backgrounds, but she has also spent a lot of time studying high achievers in elite spaces.

Maybe privileged people are attempting to stretch one of their own fixations over a problem that's too big for it to fit. Recall that the first-ever use of the term in this context referred not to a struggling low-income student but to a British poet. Some of that is going on today, too. Grit is a prominent component of twenty-first-century self-help narratives. A (very) partial sampling from Amazon: *Old School Grit: Times May Change, but the Rules for Success Never Do* (that's part of the *Sports for the Soul* series), *Grit for Kids: 16 Top Steps for Developing Grit, Passion, Willpower, and Perseverance in Kids for Self-Confidence and a Successful Life*, *The Grit Guide for Teens: A Workbook to Help You Build Perseverance, Self-Control, and a Growth Mindset*, and *Grit and Grace: Uncommon Wisdom for Inspiring Leaders Designed to Make You Think*.

These books are not marketed to low-income people. They are marketed to the sorts of people who buy social-science-flavored self-help books: mostly middle- and upper-middle-class strivers eager to leverage the latest research findings to boost their own odds of career success and fulfillment. Many of these titles offer a patina of science that takes grit's boldest claims at face value—"Psychologists tell us that the secret to a successful and happy life, more than anything else, is something called GRIT," reads the publicity material for *Old School Grit*[68]—and they generally come across as targeted at readers who might have watched Duckworth's TED Talk, or read a magazine article about her, and who are eager for more information about how to personally benefit, or help their kids benefit, from her seemingly breakthrough insight.

Self-help always sells in America, but what's interesting here is the extent to which grit fits a very important, very of-the-moment fear: that kids aren't going to do as well as their parents. This is a perennial sort of fear, of course, as evidenced by those middle schoolers selling copies of the newspaper *Grit* to develop some, but it's particularly salient and rational at the moment.

After all, my generation, the millennials, will likely be less prosperous than our parents were. Evidence has steadily piled up during the decade and a half since the modern incarnation of grit hit the scene: our home-ownership rate is lower than our parents' was at the same age,[69] we have only half the wealth our parents did at the same age,[70] and at least some of our different family-structure habits, like the older ages at which we pair off and have kids relative to earlier generations, are partially driven by economic uncertainty. Many of us entered the job market at a time when the world was bucking and convulsing, mired in the single biggest economic crisis since the Great Depression. And that was all before coronavirus sparked an economic crisis that may immiserate many people who had, to this point, been lucky enough to avoid the economic challenges facing so many Americans.

What's the solution? Maybe there isn't one. We wouldn't be the first generation punished by forces beyond our control. But that's a very unsatisfying answer, and not a very marketable one. A better, more hopeful answer is this: even if things *out there* don't improve anytime soon, there are traits *in here* we can cultivate in ourselves—or our kids can cultivate in themselves—to help us hop back on the upward-mobility ladder.

Many of us, after all, were raised during the self-esteem movement. We were told, over and over again, that we were special, that we could get whatever we wanted. As we race toward forty, that's looking less and less likely. We've been slamming our palms against a locked front door for years, wondering what we could possibly be doing wrong. If grit *did* offer a novel and useful way of measuring personal potential—if that claim applied to the world at large rather than just certain narrowly constrained settings—then it would offer a side door. It wouldn't be that we aren't smart or talented enough to succeed in a society that is becoming ever more cutthroat, but rather that we were focusing on the wrong thing.

The world is a big and scary place, and imposing structures often circumscribe our room to maneuver. That's why the reductive storytelling of grit is so appealing; that's why we turn to figures like Martin Seligman for positive thinking that can shield us from trauma, or Amy Chua for parenting advice, or Angela Duckworth for the secret to grittiness. There's always that idea of the side entrance; there's always something you, the individual, can do to regain control in a world that sometimes seems hell-bent on robbing you of it.

# 6

# THE BIAS TEST

In April 2018, two young black men were arrested at a Philadelphia Starbucks. It was an ugly story: The men were waiting for an acquaintance, who was white, to arrive. A manager, also white, asked them to leave because they hadn't ordered anything, and called the police when they refused to do so. The police arrived, there was some discussion, and ultimately the two men were led away in handcuffs as onlookers exclaimed their outrage.[1]

The incident sparked an immediate uproar over what many took to be an act of unfair racial profiling and over-policing. Quickly, the responsible manager was fired. But it didn't take long for everyone, Starbucks included, to identify a likely root cause of the incident:

implicit bias, or the unconscious prejudices that shape our attitudes and behavior without our knowledge. Soon after news of the event went viral, Starbucks announced it would be shutting down all its American stores to conduct a day of diversity training premised largely on fighting implicit bias. The event perfectly captured the fact that in the twenty-first century there is no more effective way to broadcast one's seriousness about tackling racism than by invoking this concept.

Implicit bias has enjoyed blockbuster success because there is a simple test that anyone can take to measure one's own level of this affliction: the implicit association test, or IAT. If you've been in a diversity training anytime in the last few years, it's likely you've come across this tool, which is promoted by Harvard University and a veritable army of well-credentialed social psychologists who study bias and discrimination. You can go to Harvard's Project Implicit website at implicit.harvard.edu to take an IAT yourself, and if you do you'll see that the setup is fairly simple. First, you're instructed to hit *i* when you see a "good" term like "pleasant," or to hit *e* when you see a "bad" one like "tragedy." Then hit *i* when you see a black face, and hit *e* when you see a white one. Easy enough, but soon things get slightly more complex. Hit *i* when you see a good word or an image of a black person, and *e* when you see a bad word or an image of a white person. Then the categories flip to black/bad and white/good. As you peck away at the keyboard, the computer measures your reaction times, which it plugs into an algorithm. That algorithm, in turn, generates your score.

If you were quicker to associate good words with white faces than good words with black faces, and/or slower to associate bad words with white faces than bad words with black ones, then the test will report that you have a slight, moderate, or strong "preference for white faces over black faces," or some similar language. You might also find you have an anti*white* bias, though that is significantly less common. By the scoring conventions of the test, positive scores indicate bias against the out-group, while negative ones indicate bias against the in-group.

Underpinning this exercise is a line of cognitive research which suggests that, as humans, we have an easier time connecting concepts that are already tightly linked in our brains, and a tougher time connecting concepts that aren't. The longer it takes to connect "black" and "good" relative to "white" and "good," the thinking goes, the more you are in possession of unconscious biases favoring white people over black people. By now there are many flavors of the test, including IATs that can measure ostensible unconscious prejudices against women, overweight people, or minority groups around the world; German researchers, for example, have developed an IAT that compares reaction times to traditionally German names versus traditionally Turkish ones, Turks being Germany's largest ethnic minority and the victims of generations of prejudice there. The conceit of all of these tests is that they measure something of real-world import: the higher you score on a given IAT, the more likely you are to discriminate, in subtle but sometimes important ways, against the out-group members in question.

The black-white IAT, often called the race IAT, is far and away the most discussed version of the test. That discussion began when the IAT was first unveiled in 1998 by its founders, Mahzarin Banaji, a highly regarded social psychologist who is currently at Harvard University, and Anthony Greenwald, then and now an esteemed social psychologist at the University of Washington. As the accompanying UW press release indicated, the test was advertised as a potentially powerful antiracist intervention from the start: "The same test that reveals these roots of prejudice has the potential to let people learn more about and perhaps overcome these disturbing inclinations."[2]

According to its creators and advocates, the implicit association test helps solve a mystery about America's recent racial history. If you take people at their word, as far as polling about racial attitudes is concerned, most Americans are treating one another with something like equality. And yet, more than half a century after the end of Jim Crow, all sorts of stark racial disparities persist in domains ranging from housing to education to health care. The killing of George Floyd in May 2020 cast a harsh light on racial inequities in police work, sparking nationwide protests and demonstrating, once again, that a relative dearth of explicit racism is not, on its own, any guarantee of equality.

One way to explain this situation is to posit the existence of some sort of dark matter subtly influencing interracial interactions in America. In the view of IAT proponents, that dark matter is implicit prejudice: They believe that IAT results suggest that having addressed many of the most outrageous and explicit forms of discrimination, our progress toward genuine racial equality may be continually stalled or undone by implicit bias. That is, if people who don't *feel* as if they discriminate do, in fact, discriminate, their actions could help generate many disparate outcomes. Maybe some white cops who claim racial empathy are still, deep down, more likely to pull the trigger in an ambiguous situation involving a black suspect than a white one. Maybe white real estate agents who were proud two-time Obama voters conjure up thin excuses—excuses that *feel* legitimate to them—to avoid renting nice units to black families. The data produced by the IAT, the test's creators long argued, suggest that a solid majority of Americans hold implicit biases against marginalized groups, and that therefore they are more susceptible to committing acts of bias against these groups than they might think.

The IAT wasn't the first tool researchers developed to attempt to measure implicit bias, but it was the first to fully catch on. From the start, the psychology establishment seemed to realize it could have a juggernaut on its hands: a 2001 article in the American Psychological Society's *APS Observer* magazine, for example, described the IAT as "a revolution in social psychology."[3]

The public immediately took notice as well. In the twenty-plus years since the IAT's first appearance in Seattle, it has been mostly treated as a revelatory technological innovation, garnering overwhelmingly positive media coverage. The test was covered with something like awe in Malcolm Gladwell's bestselling *Blink: The Power of Thinking Without Thinking* ("The IAT is more than just an abstract measure of attitudes," he wrote. "It's a powerful predictor of how we act in certain kinds of spontaneous situations"[4]) and in the NPR social science correspondent Shankar Vedantam's book *The Hidden Brain*, which launched a very successful podcast of the same name. The *New York Times* columnist Nicholas Kristof is a big fan and has mentioned the test on multiple occasions. "It's sobering to discover that whatever you believe intellectually, you're biased about race, gender, age or disability," he wrote in 2015.[5]

Banaji and Greenwald touted the test's real-world utility with confidence. Sure, you take the IAT sitting at a computer, engaging in a simulated task. But that simulated task, they long argued, reveals something about your propensity for *real-world behavior.* In their bestselling 2013 book, *Blindspot: Hidden Biases of Good People*, they write, "The automatic White preference expressed on the Race IAT is now established as signaling discriminatory behavior. It predicts discriminatory behavior even among research participants who earnestly (and, we believe, honestly) espouse egalitarian beliefs. That last statement may sound like a self-contradiction, but it's an empirical truth. Among research participants who describe themselves as racially egalitarian, the Race IAT has been shown, reliably and repeatedly, to predict discriminatory behavior that was observed in the research."[6]

Long before they wrote the book, both co-creators implied or stated, repeatedly, that the IAT could go a long way toward solving all sorts of societal problems. In a talk she gave at the American Psychological Society's annual convention, Banaji likened the invention of the IAT to that of the telescope: it had ushered in a revolution in how we see the world. And in a March 2000 episode of *Dateline*, Greenwald noted that "if a police officer is going to shoot two-tenths of a second faster at an African American than a European American, well, that could be a matter of life and death."[7] The implication was clear: the IAT can predict many of the behaviors that create a racially unjust society. In fact, as *Blindspot* puts it, "given the relatively small proportion of people who are overtly prejudiced and how clearly it is established that automatic race preference [as measured by the IAT] predicts discrimination, it is reasonable to conclude not only that implicit bias is a cause of Black disadvantage but also that it plausibly plays a greater role than does explicit bias in explaining the discrimination that contributes to Black disadvantage."[8]

The test's appeal owes something to an uncomfortable, provocative promise—to reveal things the test taker would perhaps prefer not to know about him- or herself. Many people get bad news when they take an IAT. One commonly cited figure stemming from the voluminous data the test's creators have collected from Project Implicit is that more than 75 percent of people score positively on the race IAT, ostensibly indicating antiblack bias.[9] (The average score is higher among white test takers, but a solid percentage of black ones receive positive scores, too.)

So the IAT spread not solely because of its scientific claims about being able to predict behavior but also because it promised an emotionally potent moment of computer-assisted introspection. Those moments began even before the test was released: the press release announcing the IAT's unveiling, for example, notes that "Banaji and Greenwald admitted being surprised and troubled by their own test results." In 2005, Banaji would go further, telling *The Washington Post* she was "deeply embarrassed" at her positive result. "I was humbled in a way that few experiences in my life have humbled me."[10] In *Blindspot* Greenwald described his first IAT session as a "moment of jarring self-insight . . . I can't say if I was more personally distressed or scientifically elated to discover something inside my head that I had no previous knowledge of."[11]

They're not the only ones to report being deeply affected by their

IAT results. Describing the emotional nature of an IAT encounter became so common that in 2008 John Tierney mused on the *New York Times* website, "It's something of a custom, when discussing the IAT, to disclose your own score on the test along with your unease."[12] In other instances, people from minority backgrounds were shocked to learn they were biased against their own group and naturally responded to this news with discomfort. In *Blink*, for example, Malcolm Gladwell describes his unease at finding out he has a moderate level of antiblack implicit bias, despite being biracial.[13] And in 2015, the radio station KQED ran a story in which a pharmacy resident of Iranian descent said she was disturbed by her IAT result. "It was like, actually, you're biased and you don't like brown people and you don't like Muslims," she told the reporter April Dembosky. "Which is interesting for me because that's kind of the two things that I am."[14]

The fact that the IAT has been marketed as not only scientifically but *emotionally* powerful, and therefore as a potent educational and self-help tool, can partly explain the enthusiasm with which it has been adopted by progressive-minded corporations as part of their diversity-training regiments. Companies ranging from old-school banks and defense contractors to Google[15] and Facebook[16]—and Starbucks, of course—have embraced the IAT as an important tool for educating employees about implicit bias and how it can lead to discriminatory hiring and workplace climates. Many universities have used the tool. And so have many law-enforcement organizations, often via consulting companies like Fair and Impartial Policing. That firm advertises on its website that it has provided implicit bias trainings to countless state, federal, and local law-enforcement agencies, from the Albuquerque and Wichita police departments to the California Department of Justice and the Massachusetts State Police Academy.[17] Through the consulting arm of their Project Implicit organization, Banaji and Greenwald are frequently hired to conduct implicit bias trainings, too, as is what appears to be a growing cohort of other IAT experts.

In light of these facts, it may feel safe to assume that the test has some serious empirical credibility. Surely it measures implicit bias with reasonable accuracy. Surely there's evidence showing that this test really does predict discriminatory behavior strongly enough to warrant all this attention.

BUT EVER SINCE THE IAT first caught on, there have been skeptics. While most of academic psychology and mainstream journalism accepted the IAT and its ramifications at more or less face value, those skeptics refused to do so. What ensued was an important academic battle that seriously dented, if not shattered entirely, the IAT's most important claims.

This battle was fought most fiercely, albeit not exclusively, between a core set of IAT proponents and a core set of critics. The proponents included Banaji and Greenwald, the creators of the test, as well as John Jost of New York University and Brian Nosek of the University of Virginia (who has also carved out his own niche as a leading advocate for reforming psychological science through study replications and other reforms, and who, as we will see, also co-authored an important meta-analysis calling the IAT's utility into question). The critics included the influential Wharton School professor Philip Tetlock, who's perhaps best known for studying why some people are better at making predictions than others, Hart Blanton of Texas A&M (a psychology methods expert), Gregory Mitchell of the University of Virginia School of Law, Fred Oswald of Rice University, Hal Arkes of Ohio State University, and James Jaccard of NYU.

It might be worth taking a step back here. How does one even measure whether a test like the IAT works as advertised? Psychometrics, or the branch of psychology concerned with creating and evaluating psychological instruments measuring anxiety, depression, implicit bias, or whatever else, offers basic guidelines, and they involve statistical benchmarks measuring how good a given instrument is. The two most important such benchmarks measure a test's *reliability*—that is, its level of measurement error (every test has some)—and its *validity*, or the extent to which it is measuring what it claims to be measuring. A good instrument needs to score solidly in both departments to be accepted by the psychological community.

Part of the critics' argument was that the IAT has serious issues on both fronts. Take the concept of test-retest reliability, which measures the extent to which a given instrument will produce similar results if you take it, wait a bit, and then take it again. This is one of the first things a psychologist will look at when deciding whether to use a given tool. If a depression test tells people they're severely depressed and at risk of suicidal ideation at noon, but free from all depressive symptoms a couple hours later, that's not a useful test. Test-retest reliability is expressed with a variable known as $r$, which ranges from 0 to 1. To gloss over some of the statistical details, $r = 1$ means that if a given test is administered multiple times to the same group of people, it will rank them in exactly the same order every time. At the other end of the spectrum, when $r = 0$, the ranking shifts every time the test is administered, completely at random. Overall, the closer you get to $r = 0$, the more the instrument in question resembles a random-number generator rather than a useful tool.

It's safe to say that most laypeople who are familiar with the IAT suppose that it provides useful information based on a single session, in large part because that's what the test's advocates have claimed for so long, and because that is almost always how it is administered. The limited available data suggest otherwise. In general, researchers are comfortable with a psychometric instrument if it has a test-retest reliability of about $r = 0.8$ or higher. But when you lump together all the different IATs, from race to disability to gender, they have an average test-retest reliability of about $r = 0.55$. By the normal standards of psychology, this puts the IAT well below the threshold of usefulness in real-world settings. There's a surprising dearth of published information on the specific test-retest reliability of the race IAT, but the individual results that have been published suggest that that figure is even lower. Depending on the study and the context, results have ranged from $r = 0.32$ to $r = 0.65$.[18] Calvin Lai, a Washington University in St. Louis professor who was previously a Harvard postdoc and the director of research at Project Implicit, ran the numbers from some of his own data for me back when I was first reporting on the test, and came up with similar results: he said he'd estimate the race IAT has a test-retest reliability somewhere in the neighborhood of $r = 0.42$. When I checked back with him in July of 2020, he pointed me toward newer literature which upped his estimate to roughly $r = 0.5$.[19]

The takeaway here is as simple as it is surprising: there doesn't appear to be *any* published evidence that the race IAT has a test-retest reliability that would be seen as acceptable for real-world use in most contexts. If you take the test today, and then take it again tomorrow—or even in just a few hours—there's a solid chance you'll get a very different result. (It should be said that there are still certain consistent patterns: most white people, for example, do score positively on the black-white IAT, ostensibly signaling antiblack implicit bias.)

As for the IAT's validity, over and over the test's proponents have made confident statements about its ability to predict behavior. In the quotation from *Blindspot* excerpted earlier, for example, Banaji and Greenwald explicitly claimed that the test does a better job predicting behavior than explicit measures like feelings thermometers in which people numerically "rate" their feelings toward different groups.[20] This is an absolutely crucial claim, and much of the IAT's cultural and academic gravitas flows directly from it. If the IAT can't predict discriminatory behavior more accurately than explicit measures of attitude can, then it's a lot less useful and interesting than its proponents have made it out to be. A major conceit of the test, after all, is that it reveals the *hidden* biases that can pop up in people who explicitly *renounce* discriminatory beliefs or intent. (A side note I've always found interesting: In 1969, a paper published in the

*Journal of Social Issues* reviewed the literature and found "a relatively low relationship . . . between measured attitudes and overt behavior" in published studies. In one study, for example, "only two of the 34 subjects said they were not willing to pose with a Negro," but twelve refused to sign a release after being photographed with one.[21] Similarly low correlations were found between people's attitudes about their work and their actual work performance, and in other domains, too. This sparked a long-running methodological debate within psychology, but the key takeaway is that with the exception of trivial counterexamples—Republican partisanship indeed correlates with voting for Republican candidates—people's stated attitudes and beliefs often tell you less about their behavior than you might think.)

In statistical terms, the architects of the IAT claimed, for a long time, that there is a meaningful correlation between two variables: someone's IAT score and how biased they act in intergroup settings. It's not easy to measure discriminatory behavior in a lab context, but researchers have come up with a variety of methods for attempting to do so. You can see whether a white participant interacts differently with a white as compared with a black experimenter, for example. Or you can have them complete a simulated hiring task in which they choose whom to "interview" by comparing résumés that are effectively identical, except that some have stereotypically white names and others have stereotypically black names. These methods aren't perfect, and many of them have been critiqued on various grounds, but that's the toolbox social psychologists are working with.

IAT researchers have published studies that they claim show reasonably strong links between IAT scores and behavior related to race and other issues, and there are plenty of these papers floating around; if you do an online search, you can find IAT papers correlating IAT scores with all sorts of important behavioral outcomes. But to reiterate: When it comes to establishing social scientific claims, interesting-seeming one-off studies get you only so far. Meta-analyses, while by no means perfect, tend to be better and tend to give more accurate answers than handpicked collections of studies. And meta-analyses have revealed that the IAT does not accurately predict real-world discriminatory behavior.

We know this because of a long, bruising meta-analytical fight that played out in the pages of the *Journal of Personality and Social Psychology*, a flagship publication for research psychologists. Starting in 2009, a team of the IAT's architects[22]—Greenwald, Nosek, Banaji, and others, with different names on different papers—duked it out with some of the test's leading critics:[23] Oswald, Mitchell, Blanton, Jaccard, and Tetlock. The arguments and rejoinders and subarguments and subrejoinders got rather complicated and technical, entailed competing meta-analyses and debates over which studies should be included, and took years to unfurl. But the result is that both critics and proponents of the IAT now agree that the statistical evidence is simply too lacking for the test to be used to predict individual behavior. The proponents conceded this in 2015: The psychometric issues with race and ethnicity IATs, Greenwald, Banaji, and Nosek wrote in one of their responses to the Oswald team's work, "render them problematic to use to classify persons as likely to engage in discrimination."[24] In that same paper, they noted that "attempts to diagnostically use such measures for individuals risk undesirably high rates of erroneous classifications." In other words, you can't use the IAT to tell individuals how likely they are to commit acts of implicit bias.

To Blanton, this was something of a smoking gun. "This concession undermines the entire premise of their web page," he told me. "Their web page delivers psychological diagnoses that even they now admit are too filled with error to be meaningful."[25] Now, there does appear to be a statistically significant correlation between IAT scores and behavior observed in studies; it's just so small as to likely be meaningless in the real world. The best and most recent careful estimate, from a key meta-analysis co-authored by Lai, Nosek, and others in the *Journal of Personality and Social Psychology*, suggests that

IAT scores account for a mere 1 percent or so of the variance in observed behavior in lab studies.[26] Viewed from one angle, this can be seen as an example of scientific progress; the authors of an important idea backed off their most enthusiastic claims as more evidence came in. The problem is that after conceding the test lacked much individual predictive validity, Greenwald continued to claim otherwise in interviews. In 2017 he told the journalist Olivia Goldhill, "The IAT can be used to select people who would be less likely than others to engage in discriminatory behavior."[27] In a 2018 email to another journalist, my friend Katie Herzog, he allowed that *interventions* geared at using the concept of implicit bias to reduce discrimination don't appear to be effective, but he also said, "The Implicit Association Test is a valluable [*sic*] educational device to allow people to discover their own implicit biases." This language certainly suggests the test is useful at the individual level.

The sometimes mixed messages surrounding the IAT has led to something of a "Schrödinger's test" situation in which it both does and doesn't predict discriminatory behavior, depending on whom you learn about it from and when. If you read Banaji and Greenwald's mass-market explanation of the test published in (bestselling) book form, you'll find out that the IAT "predicts discriminatory behavior even among research participants who earnestly (and, we believe, honestly) espouse egalitarian beliefs" and "has been shown, reliably and repeatedly," to do so. In fact, this is a "clearly . . . established" "empirical truth." You'll get the same message from some of Greenwald's quotations. If, on the other hand, you read a complicated meta-analytic back-and-forth that never got much mainstream press attention, you'll learn that the IAT isn't accurate enough to predict individual behavior. It's a frustrating example of scientists providing muddled messaging to the public.

Now, the ultimate goal of the IAT wasn't just to predict behavior, of course, but to potentially change it, and over the years researchers have developed various interventions targeting implicit bias that have taken a number of different shapes. In some, "participants either imagine how the outgroup thinks and feels, are made aware of the way the outgroup is marginalised or given new information about the outgroup, or imagine having contact with the outgroup," as one review puts it.[28] In others, "participants are exposed to exemplars that contradict the stereotype of the outgroup" or are "encouraged to activate egalitarian goals or think about multiculturalism, co-operation or tolerance." In theory, if these exercises reduce implicit bias, as measured by the IAT, such interventions could bring about concomitant reductions in biased behavior.

But on this front, too, recent efforts have proven disappointing. The aforementioned 2019 meta-analysis in the *Journal of Personality and Social Psychology* found that interventions designed to change implicit bias appeared to have a "relatively small" impact on measured IAT scores, but didn't bring with them any accompanying changes in behavior.[29] Moreover, the researchers saw some hints that even these effects might have been overstated in the literature. "Our findings suggest that changes in implicit measures are possible, but those changes do not necessarily translate into changes in explicit measures or behavior," the authors concluded. More than twenty years into the age of implicit bias, there are no proven interventions for reducing it in a manner that is relevant to real-world behavior.

THE TROUBLE WITH THE IAT starts with a notably fundamental concern: it has never been clear exactly what it measures. The tautological answer is that it measures what it measures; that is, it measures whether someone is quicker to connect positive concepts with white people and negative concepts with black people, or the reverse. But it's fair to ask this: Does an average difference in reaction time on the order of a couple hundred milliseconds constitute evidence of "implicit bias" against a certain group? This has been a source of major confusion; ever since the IAT was introduced, a great deal of media coverage and diversity-training programming has suggested that getting a high IAT score means one is implicitly biased against a minority group. The phrase "got a certain range of scores on an IAT" is often treated as synonymous with "is implicitly biased," which is in turn treated as a condition that causes someone to act in an unconsciously discriminatory way in the real world.

That isn't necessarily true, though, because all the IAT measures on its own is reaction times to different stimuli. The IAT claims to be measuring implicit bias, but that's just a claim; there's no reason to automatically accept it. What if someone who scores high on the IAT never acts in a biased manner? Can a bias be a bias if it exists only in the context of a very specific test result but never bubbles out into real-world behavior? If "implicit bias" were defined merely as "the state of having received a high score on an IAT," without that score implying a connection to real-world behavior, no one would care about implicit bias, or about the IAT.

There have always been alternative theories about what the IAT measures, though. In 2004, for example, Hal Arkes and Philip Tetlock published a paper that asked in its title "Would Jesse Jackson 'Fail' the Implicit Association Test?"[30] In it, they argued that it could be the case that people who are more *familiar* with certain negative stereotypes score higher on the IAT, whether or not they unconsciously *endorse* those stereotypes. Along those same lines, some researchers have suggested that those who empathize with out-group members, and are therefore well aware of the unfair treatment and stereotypes they are victimized by, have an easier time forming the associations that the IAT interprets as implicit bias against those groups.

In 2006, for example, Eric Luis Uhlmann, Victoria Brescoll, and Elizabeth Levy Paluck published the results of a very clever experiment they conducted on undergraduates in the *Journal of Experimental Social Psychology*.[31] The participants were assigned to either "associate the novel group Noffians with words related to oppression and the novel group Fasites with words related to privilege," or the reverse—Noffians privileged, Fasites oppressed. Then they were given a race IAT, but with Fasites and Noffians standing in for whites and blacks. As it turned out, "participants were faster to associate Noffians with 'Bad' after being conditioned to associate Noffians with oppression, victimization, and discrimination." In other words, the experimenters were able to easily induce what the IAT would interpret as "implicit bias" against a nonexistent group simply by forming an association between that group and downtroddenness in general.

Another study appeared to demonstrate a different kind of anomaly with the IAT: whites who were more concerned about *appearing* racist were scored as more "biased" by the test.[32] In the initial, since-updated version of the IAT's scoring algorithm, there was even a correlation between cognitive-processing speed and IAT score, the researchers Sam G. McFarland and Zachary Crouch found[33]: those who were a bit cognitively slower got higher IAT scores, meaning they were told they were more biased than faster test takers. Summing up this body of work, Hart Blanton and Elif Ikizer write in a book chapter, "IAT scores are contaminated by a host of non-attitudinal factors, including but not limited to a respondent's prior test-taking experience, their self-presentation motives and a range of more general skills related to intelligence, cognitive flexibility and speed."[34]

In other words, there's a significant amount of evidence that the IAT measures a variety of things apart from implicit bias itself. Which strongly suggests that many of the people who have been told they are implicitly biased over the years as a result of their IAT scores have been, in a very real sense, misdiagnosed. It's unclear why the psychological establishment should accept this when it would surely reject a similarly noisy and arguably misleading test of depression or anxiety.

Given the low rate of correlation between IAT scores and behavior, the lack of useful interventions sparked by the test, and the more fundamental, as-yet-unresolved theoretical questions about what exactly the IAT measures, it should be clear that there are serious issues with using this instrument in the real world. And yet that doesn't appear to have slowed the IAT's adoption by everyone from Starbucks to various police departments one iota.

There are reasons for that.

IN CERTAIN RESPECTS, the national conversation about race might seem to be an exception to the "age of fracture" diagnosed by the historian Daniel Rodgers. Recall that according to Rodgers the language and concepts America uses to understand itself have become increasingly individualistic and divorced from big, structural forces.

By contrast, the most prominent public intellectuals writing about race in the twenty-first century, such as Ta-Nehisi Coates, Michelle Alexander, and Nikole Hannah-Jones, offer sharp critiques of American racism that are structural in nature. Though they make different arguments and focus on different areas, all of these thinkers would deny that racism can be understood simply by "surveying the proportion of people in a society who hold 'racist' beliefs," as Eduardo Bonilla-Silva, another prominent, structurally oriented race theorist, describes a model he too disagrees with.[35] In their view, racism in America is best understood as being about not individuals but structures: the way policing and housing and schooling and so many other institutions function and bear the imprint of the past.

Hannah-Jones, for example, won the Pulitzer Prize for her introductory essay to *The New York Times*'s 1619 Project, arguing in it that "Anti-black racism runs in the very DNA of this country"—a point expanded upon in the rest of the special-edition magazine's essays, which connected structural racism to such disparate subjects as the layout of the Atlanta suburbs and the nation's lack of universal health care. Alexander wrote *The New Jim Crow*, which argued that the present-day justice system is built in a manner designed to control and oppress black people while maintaining a "color-blind" facade, and Coates's cover-story opus in *The Atlantic* made its "case for reparations" by drawing a straight, brutal line from slavery to the failure of Reconstruction to Jim Crow to the government's responsibility for urban housing segregation. These are all structural accounts. One of the most successful activist movements of the twenty-first century, Black Lives Matter, has taken a similar tack. The demand to "defund" or "abolish" the police, which became the battle cry of many protesters after George Floyd's murder, is controversial in part because it is structural. For advocates of these policies, the argument is not that individual officers are or aren't racist but that the institution needs to be reconceived in a more fundamental way. More incremental changes such as bolstering officers' de-escalation training might be better described as "institutional" rather than "structural" (there's a foggy boundary area between the two), but the same logic prevails: you're not attempting to change the individual minds of individual officers but rather changing the context in which their work takes place so that it is less likely policing will generate unjust outcomes.

As a result of this writing and activism, terms like "structural racism" have entered the lexicon, and concepts like "white supremacist" are increasingly understood to apply not just to certain individuals but, in some cases at least, to entire social structures as well. And yet, in one particular area it can be argued that race talk has grown only more individualistic in recent decades, following the path described in *Age of Fracture*: antiracism education and diversity trainings.

Today's American diversity trainings can be traced back to the 1964 Civil Rights Act, which made it illegal for all but the smallest companies to discriminate on the basis of protected categories like race, religion, and sex. The act also set up the federal Equal Employment Opportunity Commission (with state branches as well) to adjudicate alleged violations. The commission could mandate training in the wake of violations, creating a market niche for these programs.

By the 1980s, a vigorous diversity-training industry had emerged, spurred on mostly by two desires on the part of firms: first, they believed these trainings could help shield them against litigation; and second, many began to believe that more inclusive workplaces could be good for the bottom line, particularly after a famous and widely read 1987 report called *Workforce 2000* argued that the workforce was set to change rapidly during the waning years of the twentieth century. "*The workforce will grow slowly, becoming older, more female, and more disadvantaged,*" predicted the report. "Only 15 percent of the new entrants to the labor force over the next 13 years will be native white males, compared to 47 percent in that category today."[36]

These programs have always taken on a variety of forms. They have ranged from simple lists of dos and don'ts geared solely at maintaining compliance with the law to "an 'in your face,' 'admit your guilt' session for White men to 'confess and repent,'" write two diversity-industry veterans in the *Academy of Management Learning and Education*.[37] Even before diversity trainings caught on as a household concept, the quest for racial justice had melded with individual consciousness-raising in some colorful ways. The Esalen Institute, for example, held a late-1960s interracial "encounter session" advertised as follows:

> Racial segregation exists among people with divided selves. A person who is alien to some part of himself is invariably separated from anyone who represents that alien part. The historic effort to integrate black man and white has involved us all in a vast working out of our divided human nature. Racial confrontation can be an example for all kinds of human encounter. When it goes deep enough—past superficial niceties and role-playing—it can be a vehicle for transcendental experience.[38]

Let's be open-minded here and grant Esalen the benefit of the doubt: perhaps segregation is caused, in part, by people who suffer from a "divided self." Maybe a black guy and a white guy can have a powerful weekend together at Esalen; maybe it'll even be a "transcendental experience." The problem, of course, is that after it ends, they are likely returning to everyday lives lived under very different circumstances, newly reconciled selves or no. (And one of them is significantly more likely to be pulled over driving home on the Pacific Coast Highway.)

Recent experience suggests that changing attitudes, on their own, is at best a partial solution. Individual attitudes and social structures reinforce each other, of course. But twenty-first-century segregation, for example, involves a complicated set of interlocking structures that transcend any individual's beliefs: those structures keep black children in inferior schools, hamper their families' ability to move out of urban cores and into the suburbs, and inflict other harms.

Before I entered kindergarten, my parents moved our family into a bigger house about a mile away. The move was largely motivated by their desire to place my brothers and me in Newton's public schools rather than Boston's. Newton's schools are among the best in the country; Boston's are not. Newton, which is far whiter, also has more and better parks and better amenities of all sorts for a young family than large swaths of Boston. Is it possible that my parents were motivated in part by discriminatory attitudes, implicit or otherwise? Surely. But it's also worth noting that the layout of Boston and its suburbs was shaped by racism, giving rise to a system in which it is perfectly "rational" to want your children schooled in Newton rather than Boston. You don't really need discriminatory attitudes to explain why my parents and so many others made their decision; the choice was baked into the structure of the city, as determined by yesterday's (often rather explicitly racist) public policy decisions.

The most obvious response to this is that it's hard to see how structures can change if at least certain people's attitudes *toward those structures* don't change as well. If people are against a policy, they won't vote for that policy or agitate for their elected leaders to enact it. Yes, but the precise mechanisms here are complicated. There is a difference between holding an attitude ("Police brutality is out of control") and actively engaging in behavior designed to change things ("I am going to take to the streets to protest police brutality," or "I am going to form a group to show up at every city council meeting and demand the local police department adopt body cameras"). Setting aside the immensely complicated questions of which reforms should be pursued, the concept of preference intensity matters a great deal here. There's a difference between scoring high on some attitudinal measure of antiracism and actually engaging in behavior that might ameliorate racial disparities. A group of a hundred people might feel some strong moral outrage about, say, lax gun control, but they can be politically bested by fifteen well-organized ardent pro-gun activists who actually engage in real-life political activism. Remember: attitudes and behavior are different things and often aren't as tightly correlated as we think. So while attitudes can, in complicated ways, set the stage for social change—if it were still 1940 and vast swaths of white America viewed black people as fundamentally inferior, then of course any nationwide conversation about police brutality as racially forthright as the one we are presently having would be dead on arrival—in the absence of specific political action they are not enough. In 2020, the protesters who filled the streets in the wake of Floyd's murder created excitement and a sense of possibility. So too did the strong support they've gotten from fellow Americans.[39] But there are still serious questions about whether such activism will lead to concrete change, and the answer is likely to depend far more on the workings of city councils and state legislatures and the federal government than changes in individual attitudes—even if, again, there are complicated connections between the two realms.

Despite its clear limitations, the idea that raising individual white people up out of their ignorance is a key ingredient to fixing America's race problems has only become more and more popular in training settings, especially from the 1980s on. These approaches have not, in general, taken the form of getting white people to take specific political actions. Rather, they have leaned heavily on white introspection—and sometimes self-flagellation—and mastery of ever-evolving race-specific codes of behavior and etiquette. And they have always had their detractors. "Casting interracial problems as issues of etiquette put a premium on superficial symbols of good intentions and good motivations as well as on style and appearance rather than on the substance of change," argues Elisabeth Lasch-Quinn in *Race Experts: How Racial Etiquette, Sensitivity Training, and New Age Therapy Hijacked the Civil Rights Revolution.*[40] In the view of Lasch-Quinn and other critics of these programs, the stuff that made the civil rights movement *work*—sweeping arguments about all humans being part of the same extended family and deserving of dignity paired with meaningful on-the-ground activism—was largely tossed aside in favor of, well, racial etiquette, sensitivity training, and New Age therapy, much of it centered on individual white people and their beliefs and behavior.

Lasch-Quinn's book argues in part that diversity trainings and antiracism education have evolved toward ever more microscopic examinations of white people's behavior and attitudes and etiquette, often with a focus on forcing whites to experience a visceral, if not painful, reckoning with their perch atop the racial hierarchy, and often to the exclusion of other, more meaningful forms of education and activism. In her telling, by the end of the twentieth century diversity trainings and antiracist education had come to be more and more focused on individual psyches—particularly the psyches of white people. As she writes, "Psychology, at least in the form of quasi-psychological simplifications from pop psychology, has actually been the *primary* lens through which race has been understood in America in the late twentieth century."[41] This brand of psychologizing often posits a certain pathology possessed by whites that with enough striving—or training, always available for purchase—they can overcome. Lasch-Quinn was far from the first to criticize this approach. "The habit of considering racism as a mental quirk, as a psychological flaw, must be abandoned," argued the anticolonialist revolutionary psychiatrist and radical philosopher Frantz Fanon long before the era of diversity trainings.[42]

Lasch-Quinn's book came out in 2001, too early for it to discuss the IAT or other recent innovations in antiracist education, but the trend she posits has continued, if not accelerated, since then. Consider *White Fragility: Why It's So Hard for White People to Talk About Racism*, the bestselling 2018 book by the academic and popular diversity trainer Robin DiAngelo, which shot to the very top of the bestseller lists during the nationwide protests following Floyd's killing. While her book is based on a worthwhile sociological observation—that many white people exhibit discomfort and defensiveness when the subject of race or racism comes up—it is strikingly individualistic. "Naming white supremacy changes the conversation in two key ways," writes DiAngelo. "It makes the system visible and shifts the locus of change onto white people, where it belongs"—that is, *not* on the big institutions and forces traditionally seen as playing the heaviest role in generating social outcomes, but rather on individuals.[43]

Her book is, in large part, a call for white people to fight white supremacy by adopting different codes of behavior. Toward that end, she offers myriad examples of how individual white behavior stymies progress. Some are obvious and unobjectionable: it certainly doesn't help interracial interactions if, for example, white people become genuinely hostile and aggressive when the subject of race comes up, or when it is suggested that some of their success may be due to racial privilege, making it difficult to have a frank discussion. But at other points the code of behavior gets very specific and, in some cases, hard to justify. DiAngelo argues forcefully, for example, that it's important for white women not to cry when her trainings get intense if there are black people present. "There is a long historical backdrop of black men being tortured and murdered because of a white woman's distress, and we white women bring these histories with us," she writes. "Our tears trigger the terrorism of this history, particularly for African Americans."[44] Elsewhere she points out that when *she* starts crying in a cross-racial training setting, "I try to cry quietly so that I don't take up more space, and if people rush to comfort me, I do not accept the comfort."[45]

This isn't just an individualistic, behavioral approach to fighting racism but in many senses a *micro*behavioral approach: DiAngelo's program involves a careful monitoring not only of certain acts and utterances that almost everyone agrees to be offensive but also of various acts and utterances that *could be seen* as offensive if interpreted in an idiosyncratic, politically freighted manner. If a white person mentions that he or she marched in the civil rights demonstrations of the 1960s, DiAngelo advises, that comes across as an endorsement of the ignorantly offensive belief that "racism is uncomplicated and unchanging."[46] (It is unclear why it is assumed the listener would make this assumption.)

Other approaches, too, seek to tweak white people's behavior and to open their eyes to the harm they are causing unwittingly. So-called microaggressions, or "indirect, subtle, or unintentional discrimination against members of a marginalized group such as a racial or ethnic minority,"[47] are an increasingly popular diversity-training concept, particularly on college campuses. Based primarily on the work of the Columbia University professor of psychology and education Derald Wing Sue, microaggression trainings, like DiAngelo's approach, offer long lists of behaviors and utterances that might not seem offensive at first blush but that contribute to white supremacy by creating a hostile environment for members of marginalized groups. If a white person says that "America is a melting pot" or "America is the land of opportunity," she is part of the problem, because these utterances erase the importance of race and the racist nature of the United States, creating a hostile environment for people of color and even potentially causing them to become *suicidal*. (In 2017, the late clinical psychologist Scott Lilienfeld published an important rebuttal to some of the psychological claims embedded in microaggression trainings, including this one.[48]) One as-yet-unanswered challenge to microaggression theory is that the overwhelming majority of blacks and Latinos report, in polling, that they do not find many of the utterances listed as microaggressions to be offensive.[49] If an Ivy League professor who isn't black determines that "America is a melting pot" is offensive to black people but black people disagree by an overwhelming margin—77 percent of them found it inoffensive, to be exact—whose account should win out?

In New York City, a controversy surrounding the city's schools chancellor, Richard Carranza, indicates how far in this direction contemporary diversity training can go. Carranza has come under fire, particularly from conservatives, for a mandatory training curriculum for Department of Education employees that takes a very detailed approach to laying out which individual beliefs and attitudes are and aren't acceptable among school administrators. One slide that got leaked to the *New York Post* offered the following examples of "White Supremacy Culture": "Sense of Urgency," "Worship of the Written Word," "Individualism," and "Objectivity."[50]

These sorts of trainings often induce uncomfortable or emotionally intense feelings in white participants. But at core, they offer what could be considered a reassuring promise: here are the tools to dispel the thoughts and behaviors that (the theory goes) fuel white supremacy, without sacrificing anything other than your own comfort, and only fleetingly at that. You can return from Esalen (as it were) a changed person, even if you're not actually doing anything to improve the justice system or your local segregated school or your office's hiring practices. And you can return with a good story about how you tried to expurgate your own racism, and about what an intense experience you had. In part because I am from a wealthy white liberal suburb and am familiar with this attitude, I would argue that privileged white liberals living in segregated communities, in particular, want to feel that they are doing *something* about the complicated, centuries-old problem of American racism—but without necessarily having to give up anything of value.

The Columbia University scholar John McWhorter described this general style as the "religion" of antiracism in a 2015 piece in *The Daily Beast*. Using the example of white privilege acknowledgments, he wrote,

> Nominally, this acknowledgment of White Privilege is couched as a prelude to activism, but in practice, the acknowledgment itself is treated as the main meal, as I have noted in this space. A typical presentation getting around lately is 11 Things White People Need to Realize About Race, where the purpose of the "acknowledgment" is couched as "moving the conversation forward." A little vague, no? More conversation? About what? Why not actually say that the purpose is policy and legislation?
>
> Because this isn't what is actually on the Antiracists' mind. The call for people to soberly "acknowledge" their White Privilege as a self-standing, totemic act is based on the same justification as acknowledging one's fundamental sinfulness is as a Christian. One is born marked by original sin; to be white is to be born with the stain of unearned privilege.[51]

One could replace "White Privilege" with "microaggressions" or "implicit bias" or whatever new innovation will be the next hot racial-etiquette subject after this book comes out, and the same, well-grounded critique would hold: this is a style of education and training that seems much more geared at some sort of internal spiritual cleansing or awakening than at spurring actions geared at improving the world out there.

To be sure, the implicit association test is designed to be a calmer and more secular experience than a Robin DiAngelo training. In the end, though, it too probes the experiences and psyches of liberal white people in a manner guaranteed to fascinate and provoke them, and it still doesn't, fundamentally, ask much of test takers. As the legal scholar Jonathan Kahn puts it, "Ten minutes in front of a video screen, and you have taken the first step to addressing your own implicit biases—an easy, pain-free, technologically mediated, individualized means to feel connected to the enterprise of addressing racism."[52]

The IAT, then, combines several long-standing trends in the trajectory of modern diversity trainings and antiracism education. It joins individualism and consciousness-raising and the microlevel analysis of behavior and belief with the appearance of hard psychological science. Here is bias, measured to the millisecond—a flaw in one's soul that can, with effort and confession, be repaired.

AFTER I WROTE A LONG ARTICLE about the shortcomings of the implicit association test in 2017, I saw some people, particularly conservatives, missing the point of my critique entirely.[53] Sometimes, after reading the piece, they would crow that implicit bias—not the IAT, but the concept of implicit bias itself—had been "debunked."

This conclusion doesn't make sense: just because an instrument doesn't measure something well doesn't mean the thing it is measuring isn't real or is unimportant (or does a broken thermometer prove that the concept of temperature doesn't matter?). There's every reason, in light of the human brain's tendency toward stereotypes and shortcuts and other forms of overextrapolation from thin slices of the world, to believe implicit bias is very real and could affect *some* outcomes. And if, in fact, implicit bias is helping determine whom real estate agents rent to, or who gets hired and who doesn't, or whom cops shoot at, these biases could mesh with preexisting structural inequalities and make it even more difficult to make society fairer and less racially discriminatory. (Think of it this way: For structural reasons, black families have more trouble than white ones saving up for a nice home in the suburbs. If even after successfully saving up for a down payment, they are stymied by implicitly biased Realtors and mortgage agents, that could be a way in which implicit bias, at the level of individual actors, contributes to a well-established form of *structural* racism.)

So arguments that implicit bias has itself been "debunked" certainly go too far; there are strong theoretical reasons, and some empirical evidence, pointing to the existence of implicit bias. The problem is that the test's proponents themselves leap too far in the other direction: they argue that implicit bias likely plays a major causal role in generating racial discrepancies in America. There has never been nearly enough evidence to justify such a stance.

As we saw earlier, Mahzarin Banaji and Anthony Greenwald ground the IAT project in a strong claim, that "it is reasonable to conclude not only that implicit bias is a cause of Black disadvantage but also that it plausibly plays a greater role than does explicit bias in explaining the discrimination that contributes to Black disadvantage."[54] In an appendix to their book, they do allow that "racially discriminatory attitudes persist in American society,"[55] and nod to some evidence for this, but the fact remains that the strongest claim for the importance of the IAT is that it allows us to explain something that couldn't otherwise be explained.

But there is a strong case to be made that Banaji and Greenwald and so many others are contriving a mystery that doesn't actually exist and that therefore doesn't require implicit bias to solve. To focus, for simplicity's sake, on the black-white dyad, a reasonable person could argue that the vast majority of the discrepancies in outcomes between these groups can be explained by two fairly well-understood factors: the wealth gap and the present-day effects of *explicit* discrimination.

It's easy to forget the shocking magnitude of the wealth gap. "At $171,000, the net worth of a typical white family is nearly ten times greater than that of a Black family ($17,150) in 2016," notes a 2020 Brookings Institution report.[56] Arguably the single biggest correlate of well-being in the United States, whatever metrics you use, is money. Were we to strip out all the details here and simply posit a society in which there are two visibly distinct groups, one with ten times as much wealth as the other, we could predict that for generations the wealthier group would fare better than the poorer one. That's because more money means more education, better housing, and access to superior resources of every kind.

This becomes even clearer when one examines how certain institutions operate. For example, when it comes to perhaps the single biggest issue of concern in the American conversation about race—law enforcement—wealth plays a profound role. As a report to the United Nations authored by the progressive criminal justice reform organization the Sentencing Project puts it, "The source of [racial] disparities [in law enforcement] is deeper and more systemic than explicit racial discrimination. The United States in effect operates two distinct criminal justice systems: one for wealthy people and another for poor people and people of color. The wealthy can access a vigorous adversary system replete with constitutional protections for defendants. Yet the experiences of poor and minority defendants within the criminal justice system often differ substantially from that model due to a number of factors, each of which contributes to the overrepresentation of such individuals in the system."[57] Again, the wealth gap alone could explain much of the racial disparity in the justice system, even in the (purely hypothetical) absence of any sort of ongoing present-day discrimination of any sort.

But of course discrimination still exists. Many real estate agents do disfavor black home buyers just as many police officers are more likely to hassle black kids on the streets, or do worse. Here's where things get particularly complicated, because when evidence of actual discrimination presents itself, it is difficult, and sometimes impossible, to discern whether that discrimination is implicit or explicit. Perhaps the strongest evidence that implicit bias contributes significantly to inequality comes from so-called audit studies. These real-world experiments show that when two identical résumés are sent out, one with a stereotypically "white" name and another with a stereotypically "black" name, the former gets more interviews—sometimes a lot more. The most famous of these studies, conducted by Marianne Bertrand and Sendhil Mullainathan in 2004, found that applicants with "white" names were 50 percent more likely to receive a callback than applicants with "black" names, a gap that was "uniform across occupation, industry, and employer size."[58] Many researchers interpret this gap as being driven significantly by implicit bias because they assume that in the twenty-first century it is unlikely a given HR manager is going to be engaging in explicit discrimination. Not everyone agrees with this interpretation; the legal scholar Michael Selmi notes that this study "is routinely identified as involving implicit bias" but asks, "Is there any reason to conclude that implicit bias explains the results?"[59] But it feels safe to say first that this sort of experimental work does prove the existence of *some* sort of bias and second that we can attribute at least *part* of that bias to unconscious influence, because it's unlikely, in 2020, to be entirely explicit.

Even here, though, things get fuzzy. For one thing, while there have been some replications of Bertrand and Mullainathan's work, the research doesn't always point in a straightforward direction. In a very clever study conducted on the Chicago labor market, Nicolas Jacquemet and Constantine Yannelis sent out three types of résumés: "one with Anglo-Saxon names, one with African-American names, and one with fictitious foreign names whose ethnic origin is unidentifiable to most Americans."[60] The Anglo-Saxon names were almost a third more likely to yield callbacks than the African American *or* ethnically unfamiliar names, which the researchers describe as "evidence that discriminatory behavior is part of a larger pattern of unequal treatment of any member of non-majority groups, *ethnic homophily*," or the human tendency to like people who are similar to us. Ethnic homophily seems a more plausible explanation of this result than implicit bias as traditionally understood, unless we assume hirers have equal levels of implicit bias against black Americans and ethnically ambiguous foreigners, which seems unlikely.

Another audit study, conducted in 2016 with tenfold the sample size of the original, found "no consistent pattern of differences in callback rates by race, unlike Bertrand and Mullainathan."[61] The data sleuth Uri Simonsohn dug into the methodologies of the two studies to try to figure out why and conducted a clever informal experiment on Amazon's Mechanical Turk, a website where users can complete simple tasks, including participating in scientific experiments, for small sums of money: He paired a "black" name from the original study with a "black" name from the new one and asked respondents to rate, on a scale, which one they believed others would see as having a higher socioeconomic status, or SES. The results suggested a class confound was at work here: In the latter study, the names were perceived as wealthier, and they weren't discriminated against. Simonsohn argues that the results of these and other studies "are as consistent with racial as with SES discrimination. The SES account parsimoniously also explains this one failure to replicate the effect." That said, this isn't a peer-reviewed finding, and Simonsohn cautions that his result should be seen as, at best, tentative, because "we are comparing studies that differ on many dimensions."[62] The point is simply that this body of literature, which on net *does* find that applicants with black-sounding names face more bias than those with white-sounding ones,[63] does not tell a consistent story about the cause of that discrimination.

If we return to law enforcement, it's worth noting that there's no shortage of evidence that explicit racism persists in that area, even if few people reveal it in survey questions (so-called social-desirability bias can artificially deflate positive responses to questions indicating racial animosity). When federal authorities investigated the Ferguson, Missouri, police department, for example, they found that "Ferguson's harmful court and police practices are due, at least in part, to *intentional* discrimination, as demonstrated by direct evidence of racial bias and stereotyping about African Americans by certain Ferguson police and municipal court officials."[64] A similar investigation of Chicago's police force found a "pattern or practice of force in violation of the Constitution," and it revealed behavior that would be hard to explain away as implicit bias: "Black youth told us that they are routinely called 'nigger,' 'animal,' or 'pieces of shit' by CPD officers . . . One officer we interviewed told us that he personally has heard coworkers and supervisors refer to black individuals as monkeys, animals, savages, and 'pieces of shit.'"[65]

Housing is another area in which it feels premature to blame ongoing disparities and discrimination primarily on implicit bias. In 2019, *Newsday* published the results of a sprawling, years-long, data-driven investigation of real estate practices on Long Island—one that "found evidence of widespread separate and unequal treatment of minority potential homebuyers and minority communities" there.[66] Another, more scientifically conducted study, published in 2013 by the Urban Institute, found that while racial bias in home renting and selling had declined over the years, "real estate agents and rental housing providers [still] recommend and show fewer available homes and apartments to minorities than equally qualified whites."[67]

It is hard to ignore the role of explicit bias at work here. It would be Pollyannaish to imagine that these two police departments and these two investigations of housing practices are distant outliers. In a country that elected a man who explicitly called for Muslims to be banned from entering the United States, it feels premature to assume, mostly on the basis of contextless survey data, that explicit bias no longer plays a major role in generating various outcomes or that its unconscious cousin is now more important.

Because of their enthusiasm for attributing various discrepancies to implicit bias, advocates for this model of understanding American racism sometimes appear to let various institutions off the hook. In her book *Biased: Uncovering the Hidden Prejudice That Shapes What We See, Think, and Do*, the MacArthur-winning Stanford psychologist Jennifer Eberhardt, who has studied and consulted extensively with police officers, reflects on the impact police work can have on individual officers: "'MALE BLACK.' 'MALE BLACK.' 'MALE BLACK.' 'MALE BLACK,'" she writes. "This is what officers in Oakland hear, booming from their police radios, hundreds of times every day." She goes on to argue that "that repetitive pairing can easily lead to an association of blackness with crime that becomes automatic, expected, routine."[68]

Certainly so. But is it likely these officers are unaware that they are spending so much time searching for and having negative interactions with darker-skinned suspected criminals? Perhaps more to the point, if we were to list all the different forces shaping a given police officer's behavior, how prominent a role would implicit bias play? There's strong reason to believe a focus on implicit bias can obscure the fact that many law-enforcement policies that lead to terrible outcomes are the result of explicit decisions made by specific people in positions of power. Eric Garner's horrific, video-recorded death by choking at the hands of an NYPD officer was sparked when he was confronted for selling "loosies"—individual cigarettes—on the streets. This was a specific policy decision made by the NYPD to enforce the law in a specific manner, and in general this sort of "quality of life" policing is less likely to descend harshly upon whiter and wealthier neighborhoods. In addition, Garner was killed by a type of police chokehold that had already been outlawed in 1993, so if the officer in question had followed the NYPD's own stated policies, the death would have likely been averted.[69]

Along similar lines, Jonathan Kahn points out that "among the key *structural* differences between European and American policing practices is the fact that in 1989 the U.S. Supreme Court deemed it constitutionally permissible for police to use deadly force when they 'reasonably' perceive imminent and grave harm," while European nations are subject to a stricter "absolutely necessary" standard.[70] This could help explain why American police shoot far more civilians than European police do—as could a range of other factors such as the number of guns on the streets in the United States and the fact that our patchwork mental-health-care system leaves more vulnerable people homeless and at risk of dangerous encounters with the police, who are often thrust into the role of untrained de facto crisis counselors. All of which is the result, at least in part, of *specific policy decisions*. To posit implicit bias as a particularly important variable here is to run the risk of erasing policy decisions that simply matter more. That's doubly true when one recognizes the weaknesses and question marks in the implicit bias literature and the shakiness of the IAT itself.

In light of all this, it's clear that we simply don't know how large a role implicit bias plays in shaping ongoing American racial discrepancies. And it shouldn't come as a surprise that many people dedicated to better understanding and addressing inequality in America do not believe implicit bias is the most promising avenue. "I think unconscious racial prejudice is real and consequential," Robb Willer, a sociologist at Stanford University, told me when I was first reporting on the IAT. "But my sense is that racial inequality in America is probably driven more by structural factors like concentrated poverty, the racial wealth gap, differential exposure to violence, the availability of early childhood education, and so on."

Given the many ways racially discriminatory outcomes can manifest themselves, the burden should be on implicit bias's evangelists to prove that it plays such an outsize role as to warrant the attention and dollars it has received. They have yet to make this case convincingly.

THE POPULARITY OF THE IAT has turned certain aspects of the conversation about race upside down: implicit bias is a major area of focus for social psychologists not because anyone has proven critics of implicit-bias-centered approaches wrong but rather because the most popular antiracist tool available happens to be concerned with implicit bias. Of course, IAT experts wouldn't be the first to have fallen for this very human tendency. "Like the proverbial boy with a hammer to whom the whole world looks like a nail, to the expert with an algorithm, the whole world of decision-making is reducible to the terms of that algorithm," notes the philosopher Conrad Brunk.[71]

Patrick Forscher is a research scientist at the Université Grenoble Alpes in France. He previously ran an intergroup relations lab at the University of Arkansas and was the lead author, with Calvin Lai, on the major 2019 IAT meta-analysis that delivered unimpressive results for the IAT. As a result of his research he has grown significantly more skeptical of the IAT over time—even going so far as to publicly say and write that he no longer uses the test or recommends its use in his consulting work.

Forscher believes the IAT discourse perpetuates itself at least in part for extrinsic political reasons. "The problem is that implicit measures, and the IAT in particular, became a critical part of a political narrative about why disparities between social groups exist in the United States," he said. "Thus, claims about implicit measures became, to a certain extent, political claims, not just scientific claims." This likely makes it harder to evaluate the test critically: when scientific ideas becomes politicized in this manner, when supporting or being critical of individual claims comes to be seen not just as a scientific act but as a political one—*Why would you deny the importance of this test measuring implicit racism, which we all know is an incredibly important thing to do?*—it's easier for half-baked ideas to flourish.

"Focusing so narrowly on implicit bias risks ignoring the complexity of the problems, like racial disparities, that are argued to be caused [by] implicit bias," said Forscher. "Any problem as tenacious and long-standing as racial disparities is unlikely to be caused by any one thing. Surely, then, it is worthwhile for psychologists interested in resolving problems like racial disparities to investigate other possible causes of disparities, such as intentional or structural processes." It's not that psychologists are entirely ignoring these other causes; it's just that the IAT, by dint of its cultural and academic resonance, has generated a strong gravity well that sucks in money and researchers. If you study the IAT, you can easily produce quantitative data, you can help promote an interesting and provocative story line about race in America, and you can be a part of one of the most popular and widely publicized contemporary approaches to solving serious racial issues.

The Starbucks story that begins this chapter shows the reach and breadth of this idea. The IAT is such a powerful cultural meme at the moment that everyone simply assumed the manager in question was acting under the influence of implicit bias. There's actually no evidence she was; there are countless other possibilities that simply went ignored, ranging from her having a bad day and lashing out at the next customers who annoyed her to her being explicitly racist. These possibilities were ignored because implicit bias is often *the way* we talk about racial incidents in America now.

BEFORE WE LEAVE IMPLICIT BIAS BEHIND, one additional point is warranted: institutions can address the possible harms of implicit bias *without measuring or attempting to alter individuals' levels of bias.* This is crucial.

Think about it this way: Let's assume, as seems reasonable in light of the data, that hiring managers have *some* propensity to react negatively to certain nonwhite-sounding names, perhaps penalizing them in effect. Let's also assume, as seems reasonable, that contextual factors might affect the strength of this impulse at a given moment (How hungry are they? How tired? How bored?) and that these contextual factors, plus the sheer complexity of implicit bias itself, plus the well-documented weaknesses of the IAT, mean that at the moment there's no good way to know how implicitly biased a given hiring manager is, let alone to help free him from any implicit bias he possesses.

Here's where the kinds of institutional reforms we discussed in the power-posing chapter warrant a reprise. In a situation where discrimination is a possibility, it *doesn't matter* that you don't know how biased individual deciders are, or that you aren't targeting their bias itself at the individual level, any more than it matters, in administering a flu vaccine, exactly how susceptible each person being inoculated is to the illness. By introducing certain reforms to the selection process, in a smart, evidence-based way, you can reduce the *overall probability of bias* while entirely sidestepping the issue of how noisy and difficult bias is to "measure" at the individual level. Recall that interviews, and particularly unstructured ones, may be worthless as a means of evaluating a prospective employee's potential and are exactly the sorts of settings in which bias—whether homophily or implicit bias or whatever else—can rear its head. So why not do away with them?

While abandoning interviews would require a potentially painful break from tradition, doing so—or at least reducing their weight in the hiring process—could, on its own, reduce the possibility of discriminatory influences creeping in. The same goes for certain bits of potentially bias-triggering pieces of information that appear on résumés and cover letters, like names: there might be some ambiguity to interpreting those audit studies, but they *did* reveal bias that tended to favor white applicants, so simply removing names from the first step of the hiring process nips in the bud the very possibility of implicit (or explicit) bias in this area.

So despite the IAT's weaknesses, there should be optimism about reducing the influence of bias in certain domains. Often it's possible to change the structure of decision-making for the better. Such interventions aren't a cure-all, they won't always work, and each one needs to be evaluated on its own merits, but the point is that what makes these ideas promising is that they are designed to address behavior in a realistic way—ignoring individuals' characteristics, in a sense, and focusing instead on how each of us is influenced by the social and professional systems in which we are embedded. Sure, you can try to train a female business student to be more confident seeming, or try to train a hiring manager to be less implicitly biased, but there's good reason to believe you'll have more success changing the former's business school so that overconfidence isn't unduly rewarded in the first place, and the latter's workplace so that implicit bias doesn't even have a chance to hijack decision-making when there's hiring to be done. These approaches deserve more attention than they get. The problem, again, is that they aren't quite as eye-catching as interventions that promise to reform individuals.

# 7

# NON-REPLICABLE

It's not a surprise that humans are fascinated by the idea that they are influenced by forces beyond their ken. Much of what underlies our own cognition is hidden from us, after all. There's a lot going on "under the hood," as it were. I can't necessarily explain why I like kale, or why I suddenly got sad the other day, or why someone I met a few weeks ago gave me a creepy first impression.

These are the questions that eventually gave rise to the study of social priming, a blockbuster concept in social psychology defined by the pseudonymous neuroscientist and science writer Neuroskeptic as "the idea that subtle cues can exert large, unconscious influences on human behaviour."[1] Social priming is one of the main reasons social psychology has generated more popular attention than just about any other area of psychology in recent years. It underpins the implicit association test and has generated many findings that have been accepted by experts as scientific truths.

Of course social psychologists weren't the first people to wonder about the possibility of unconscious influences on our behavior and preferences. Even before Freud, the topic had already been the subject of millennia of speculation and theorizing. To take one of countless examples, in *The Republic*, Plato explained (through Socrates) that the human soul could be seen as consisting of three different creatures: a human, a lion, and "one of those creatures that ancient legends say used to exist . . . [such as] The Chimera, Scylla, [or] Cerberus." To do injustice, he argued, is "to feed well and strengthen the multifarious beast, as well as the lion and everything that pertains to the lion; to starve and weaken the human being, so that he is dragged along wherever either of the other two leads." To do justice, on the other hand, is to do "what will give the inner human being the greatest mastery over the human being, to get him to take care of the many-headed beast like a farmer, feeding and domesticating the gentle heads and preventing the savage ones from growing."[2]

The idea that our behavior is subject to the control of competing forces we are unaware of, or at best dimly aware of, would prove quite durable. In 1950s America, this concept was increasingly complemented by the related idea that particular events in the external world bat us around to and fro, cognitively speaking, without our knowledge. That was also when popcorn sparked a moral panic that captivated America. Well, popcorn and Coke, to be exact.

In 1957, James Vicary, an advertising expert and social psychologist, announced that he had run a devious experiment on innocent moviegoers. His "subjects" had come to a theater in Fort Lee, New Jersey, to see *Picnic*, a film about a handsome drifter who rolls into a small Kansas town Labor Day morning and starts causing trouble (as handsome drifters are wont to do). As they watched the film, text instructing them to "eat popcorn" and "drink Coke" was flashed on-screen for intervals so brief they didn't consciously notice what had happened. The subliminal messages did affect their appetites, though; Vicary announced that as a result of his manipulation Coke sales edged up about 18 percent, while patrons gobbled 58 percent more popcorn than usual.

The nation exploded with outrage. As the psychology researcher Anthony R. Pratkanis explains in a 1992 article in the *Skeptical Inquirer* about America's recurring waves of interest in subliminal influence,[3] Vicary's stunt led to an FCC investigation and a formal prohibition on the practice from both that body and the National Association of Broadcasters, as well as a number of furious articles. "Welcome to 1984," wrote Norman Cousins in a *Saturday Review* essay sounding the alarm about subliminal advertising. "Nothing is more difficult in the modern world than to protect the privacy of the human soul," he warned.[4]

As it turns out, all the available evidence suggests Vicary's "study" was nothing of the sort: The results were never published in a full, peer-reviewed format, and there is some evidence to suggest the experiment itself was a hoax entirely.[5] But that incident, along with the success of Vance Packard's *The Hidden Persuaders*, a 1957 exposé of the cutting-edge techniques advertisers were supposedly using to manipulate consumers, highlighted the extent to which cold-war-era Americans were riven by fears of secret manipulation by malevolent forces. (Of course these fears intermingled with worries about Communists, particularly after a string of scary incidents during the Korean War in which American soldiers captured by North Korea appeared to have been brainwashed into saying terrible things about their country and its values.[6])

The specter of subliminal messaging provoked fear, but there were clear upsides if you were trying to sell something. In the ensuing decades, countless self-appointed experts would claim to have discovered the key to harnessing the power of subliminal persuasion.

Hucksters and true believers alike set up shop, trying to sell salespeople on the power of these techniques. A 1973 bestseller by Wilson Bryan Key, *Subliminal Seduction*, presented a credulous overview of the subject that treated Vicary's popcorn "study" as a true event.[7]

The idea of subliminal messaging became so deeply ingrained in the United States that it reached the pinnacle of pop culture: *The Simpsons*. In one early episode, Homer purchases a subliminal-messaging audiocassette to listen to while he sleeps that will, according to a commercial, help him control his ravenous appetite. Unbeknownst to him, he is sent a vocabulary-building tape instead. It works a little too well: When Marge asks whether he feels his appetite decreasing, the normally monosyllabic goof replies, "Ah, lamentably no. My gastronomic rapacity knows no satiety." (Here's Homer a bit later in the episode, carton of ice cream in hand, after the effects of the vocab-building tapes have worn off: "Marge, where's that . . . metal deally . . . you use to . . . dig . . . food?" Marge: "You mean a spoon?")

In another, much darker example of the intersection of subliminal messaging and pop culture, the British heavy-metal band Judas Priest was accused in a lawsuit of causing the tragic suicides of two American young men who shot themselves after listening to the band's album *Stained Class* on repeat for hours in December of 1985; one died immediately, the other three years later as a result of his injuries. The young men's families claimed that the dark two-word imperative "Do it" was sprinkled throughout the album and triggered the suicides. "Judas Priest's lead singer, Rob Halford, testified that the sound was merely his exhaling during the recording session," reported *The New York Times*, even singing one song from the album in court "to illustrate his point on the witness stand." The band won.[8] In reality, despite all these fears and despite all this excitement, subliminal advertising didn't turn out to work—at least not according to a 1996 meta-analysis that showed that it had "very little effect."[9]

That doesn't mean the idea that humans can be swayed in subtle ways by seemingly fleeting or inconsequential information is entirely bunk, though. The field of priming, which would give rise to social priming, suggests otherwise, and it's been around since about the middle of the twentieth century; call it the more scientific counterpart to the subliminal-messaging craze. The precursors to priming studies were fairly straightforward. In one study from 1950, for example, when M.I.T. students were told a visiting instructor they were about to meet was "warm," they rated him more favorably than when they were told he was "cold," despite the rest of the description they were given being identical.[10]

The concept of priming itself entered the psychological lexicon a year later, in a paper by the psychologist Karl Lashley about the seeming ease with which humans can produce strings of speech.[11] And in 1958, L. H. Storms accidentally discovered that after he gave students a list of words to memorize, those words were more likely than expected to pop up in a subsequent, unrelated free-association task.[12] Two years later, Sydney Segal and C. N. Cofer showed they could achieve a similar effect by merely *exposing* experimental subjects to the words in question rather than asking the subjects to memorize or otherwise think about them explicitly.[13] These and other key early priming studies demonstrated that even when subjects couldn't recall a given word they were exposed to or asked to memorize, it was still floating, somewhere, just out of reach, in their brains; it was still more likely to pop up during subsequent tasks than it should have been.[14]

Why do these effects exist? At root, researchers believe that priming is one of the cognitive shortcuts carved by evolution to help us navigate the world. If we had to consciously think through all the stimuli thrown at us, we'd be paralyzed by indecision and, the evolutionary thinking goes, promptly be eaten by something larger and hungrier than us. Our potential for priming exists in part because conscious cognition is *costly*. Imagine if every time you saw a dog, you had to think hard to recall exactly what animal you were looking at and what you could expect from it—*Let's see: four legs, kinda shaggy, tail wagging, likely to be friendly . . .* Instead, a web of near-automatic canine associations is activated in your brain when you see one.

The general idea that less effortful thinking can be a good thing was summed up eloquently by the British philosopher and mathematician Alfred North Whitehead in the early twentieth century, long before anyone had heard of social priming: "It is a profoundly erroneous truism, repeated by all copy-books and by eminent people when they are making speeches, that we should cultivate the habit of thinking of what we are doing. The precise opposite is the case. Civilization advances by extending the number of important operations which we can perform without thinking about them. Operations of thought are like cavalry charges in a battle—they are strictly limited in number, they require fresh horses, and must only be made at decisive moments."[15]

It's important to note that priming effects can theoretically be sparked both by stimuli we barely notice and by stimuli we don't notice at all. What's *unconscious* is the effect the stimulus has on our cognition or behavior, not necessarily the stimulus itself. Beyond priming research, it's well established that stimuli that aren't consciously noticed by people can nonetheless affect them, and even that—to oversimplify and anthropomorphize—one part of the brain can be aware of something other parts are not. Research on "blindsight" demonstrates this in a fascinating matter. In certain cases, people who are visually impaired may have no conscious experience of seeing something and yet react in a manner they could only have if they had registered the stimulus with their eyes.

While a variety of priming findings were published from the early 1950s on, it wasn't until 1977 that "social priming" arrived in the form of an important study published by E. Tory Higgins, William Rholes, and Carl Jones—a study that neatly encapsulates everything that excites social psychologists about this subject. Higgins and his colleagues exposed a group of sixty Princeton undergrads to a series of "different personality trait terms" as part of what they were told was a perception task. Then the students read the following paragraph about a man named Donald:

> Donald spent a great amount of his time in search of what he liked to call excitement. He had already climbed Mt. McKinley, shot the Colorado rapids in a kyack [*sic*], driven in a demolition derby, and piloted a jet-powered boat—without knowing very much about boats. He had risked injury, and even death, a number of times. Now he was in search of new excitement. He was thinking, perhaps, he would do some skydiving or maybe cross the Atlantic in a sailboat. By the way he acted one could readily guess that Donald was well aware of his ability to do many things well. Other than business engagements, Donald's contacts with people were rather limited. He felt he didn't really need to rely on anyone. Once Donald made up his mind to do something it was as good as done no matter how long it might take or how difficult the going might be. Only rarely did he change his mind even when it might well have been better if he had.

During the perception task, some of the students had been exposed to a set of terms that could describe Donald negatively—"reckless," "conceited," "aloof"—others to terms that could describe him positively—"adventurous," "self-confident," "independent"—and a third to terms that didn't apply to Donald and his adventures.

The researchers found that being briefly exposed to those words affected how the students evaluated the vignette about Donald, which was intentionally written to be ambiguous (and confirmed to be so by a pilot group of study participants). The subjects who were exposed to "reckless," "conceited," and "aloof" were more likely to view Donald as a blundering fool, while those exposed to "adventurous," "self-confident," and "independent" were more likely to view him as a fun-loving swashbuckler type. Though the sample sizes were quite small, of the ten students exposed to the positive terms prior to reading about Donald, seven of them viewed Donald positively, one negatively, and two in a "mixed" light. Of the ten students exposed to negative stimuli words, just one viewed Donald positively, seven negatively, and two mixed.[16] The researchers appeared to be able to profoundly affect the way a group of students interpreted an ambiguous situation simply by exposing them to certain words in an unrelated task, and the students were oblivious to the words' influence on them.

This study, wrote John Bargh and Tanya Chartrand decades later, "revealed for the first time how an individual's recent experience could affect—in a passive and unintended way—his or her perceptual interpretation of another person's behavior."[17] And it's the sort of finding that should make anyone who is curious about human nature perk up a little bit. If our beliefs about others can be so easily swayed by irrelevant stimuli, it calls many things about human social relations into question. (It's important to note that "social priming" effects sometimes have nothing to do with social life, which is why some people have argued the phenomenon should be called behavioral priming or something else. I'll stick with "social priming" because that's the most common term; if the phenomenon in question involves subtle or unconscious influence altering behavior or thought in an unconscious manner, it's social priming.)

Social priming was one of the most famous corners of psychology for perhaps two decades, starting in the mid-1990s. It exploded in stature largely due to the work of Bargh, a Yale social psychologist whose name is more or less synonymous with the subfield itself. In the 1990s, he ran a lab at New York University that greatly accelerated the popularization of social priming by producing a number of widely touted findings. Perhaps foremost among them was a 1996 study in the *Journal of Personality and Social Psychology* he co-authored with Mark Chen and Lara Burrows in which the researchers, via a scrambled-word task, exposed some participants to terms associated with the elderly like "Florida," "old," "lonely," and "wrinkle." Those who had been exposed to these words subsequently walked down a hallway slower—as measured by a sneaky confederate of the researchers—than members of a control group who hadn't been primed in that manner. "Because there were no allusions to time or speed in the stimulus materials," write the authors, "the results of the study suggest that the elderly priming stimuli activated the elderly stereotype in memory, and participants subsequently acted in ways consistent with that activated stereotype."[18]

It's easy to see why an experiment like this is so captivating. The elderly-priming result suggests subtle influences might nudge not only our judgments or decision-making but also areas of physical action we generally think of as being under our own full volitional control. The idea that simply being exposed, briefly, to words associated with the elderly might cause us to embody stereotypes associated with them and move in a more geriatric manner is surprising, exciting, and just a little scary—for many of the same reasons the idea of buying popcorn because of subliminal messaging is.

If the elderly-priming study was the initial priming experiment "hit" to fully cross over into the mainstream, it was only the first of many. Soon social priming was producing wave after wave of fascinating findings. In one, simply reminding people of money made them more amenable to free-market capitalism.[19] In another, priming students with the word "professor" caused them to perform better on a quiz.[20] Some of these studies would sound familiar to anyone who had followed the subliminal influence craze. The researchers Johan Karremans, Wolfgang Stroebe, and Jasper Claus even published one study that was fairly similar to Vicary's fake one, calling it "Beyond Vicary's Fantasies." They found that priming people with brief exposure to the brand Lipton Iced Tea—too quickly for it to be consciously noticeable—made them want that beverage more, albeit only when they were already thirsty and only for a short period.[21]

At a certain point, the field of social priming began piling up results that seemed not only impressive but astounding—maybe even miraculous. In 2011, for example, one team found that, as per their article's title, "a single exposure to the American flag shifts support toward Republicanism up to 8 months later."[22] Another study, from 2012, found that among a sample of college students, exposure to Rodin's famous statue *The Thinker* was associated with a *twenty-point lower* belief in God, on a hundred-point religiosity scale, compared with exposure to another famous statue, of an ancient Greek athlete winding up to hurl a discus.[23]

There was a period—a brief one—when some of the smartest people on the planet simultaneously believed social-priming effects *were* miraculous and that they were robust and well-founded. In 2011, the Nobel Prize–winning behavioral economist Daniel Kahneman wrote about the power of social priming in *Thinking, Fast and Slow*, his vital book on the difference between deliberative, rational cognition and cognition based more on gut impulses or hasty assessments: "When I describe priming studies to audiences, the reaction is often disbelief." A paragraph later: "The idea you should focus on, however, is that disbelief is not an option. The results are not made up, nor are they statistical flukes. You have no choice but to accept that the major conclusions of these studies are true."[24] That's how confident Kahneman was about priming research about a decade ago.

But some of these studies really do ask us, implicitly, to forget much of what we already know about human behavior and decision-making. For example, take another study about flags published in the *Proceedings of the National Academy of Sciences* in 2007.[25] In that one, the authors, Ran R. Hassin, Melissa J. Ferguson, Daniella Shidlovski, and Tamar Gross, found that when Israelis were exposed to an Israeli flag flashed for sixteen milliseconds—so quickly they didn't notice—it affected their political preferences. Subjects who initially scored high on a scale designed to measure Israeli nationalism were shifted significantly to the political left, becoming more in favor of a Palestinian state and more sanguine about Israel's then-imminent pullout from the occupied Gaza Strip, while those who scored low on that same scale were shifted in the opposite direction on those issues (as the authors note, the study was conducted "in the weeks that preceded Israel's withdrawal from Gaza, a point in time (August 2005) in which the role of the settlers in Israel's history and future was hotly debated"). The flag priming had enough of an impact that for those in the primed group, but not those in a control group that didn't see the prime, nationalistic and non-nationalistic Israelis came to express similar political opinions on these issues. And according to the researchers, the flag priming's effects reached beyond the lab, affecting participants' real-world voting behavior similarly.

As anyone who has spent time with Israelis from different parts of the political spectrum can attest, this is a remarkable claim; such differences are not easily bridged. A finding like this runs contrary to a vast body of evidence from sociology and political science and political psychology about just how difficult it is to change people's opinions when it comes to emotionally charged, highly politicized issues (issues that are new to a given person, and that don't connect viscerally with their political or personal identities, are another story; there people generally appear to be more malleable). And for the average Israeli, nothing could have been more emotionally charged than questions pertaining to Palestinian statehood and the occupied territories. It is simply hard to believe that a sixteen-millisecond exposure to an Israeli flag—a symbol many Israelis likely see in one form or another every day—could bring the beliefs of nationalistic Israelis in line with those of dovish Israelis, unless priming can swamp many *other*, preexisting, much more empirically established influences governing why people develop and hold fast to their political beliefs. It probably isn't an accident that Hassin and his colleagues' study doesn't engage with much of the past literature in political psychology and other relevant areas.

But unlikely results, if anything, only helped propel the social-priming craze forward. Priming studies were irresistible bait to many journalists precisely because they were so counterintuitive, and many fit a "This One Cool Trick Can . . ." content formula that seemed to take over the internet around 2010 or so. Priming could make us more productive, more tolerant, and just *better* in general, and could do the opposite, if we weren't careful, as in the case of implicit bias ostensibly causing white people to act in a discriminatory way when primed with stimuli associated with black people.

Naturally, priming got the most attention when it was connected to pressing societal concerns. Certain enthusiastic fans of priming speculated that the research could help the world in myriad ways. In the *Harvard Law Review*, for example, the implicit association test expert and UCLA legal scholar Jerry Kang suggested future research into the possibility of "requiring debiasing screensavers as part of a settlement in a discrimination suit," the idea being that exposure to positive exemplars of (say) black achievement could undo the mental primes that ostensibly lead to racist behavior on the part of hiring managers or judges or whoever else.[26]

Another memorable melding of social priming and social justice came in 2014, when a study published in *Proceedings of the National Academy of Sciences* purported to find that female-named hurricanes had, over the decades, killed more people than male ones, and argued that this was likely the result of implicit sexism: female names were seen as less threatening, the theory went, and thus caused people to take fewer precautions.[27] As CNN summed it up, "Apparently sexism isn't just a social problem—if you're in the path of a hurricane, gender bias might actually kill you."[28] Other outlets followed suit.

The "himmicanes" study nicely captured what is so appealing about social priming and why so many people viewed it with so little skepticism: it is an area of research that claims to hold the potential, through simple tweaks, of ameliorating many problems ostensibly caused by human susceptibility to primes and biases. And nothing better captures the scope and audacity of these claims than John Bargh's 2017 book on the subject.

---

BEFORE YOU KNOW IT: *The Unconscious Reasons We Do What We Do* is a bible for Primeworld, a realm in which it's taken for granted that primes and biases can go a long way toward explaining societal problems and how to fix them. It takes a broader, more sweeping view than other recent pop-psychology books like Angela Duckworth's *Grit* or Amy Cuddy's *Presence*. Those books tend to focus on one or two areas where their core concepts are most relevant (grit matters the most for school and professional life, while power posing matters the most for women in the workplace). *Before You Know It*, on the other hand, argues that priming has a major effect on just about everything. The framing is right there in the title: it's not that priming effects influence what we do; it's that they are *the reasons* we do what we do.

Upon completing the book, which hit shelves with glowing blurbs from Malcolm Gladwell and the superstar Harvard psychologist Daniel Gilbert, the lay reader could be excused for believing that priming is a uniquely powerful force guiding human behavior and belief. But if you read *Before You Know It* with other approaches to understanding human behavior and cognition in mind, or with some knowledge about the present, troubled state of psychological science, it quickly becomes clear that Bargh's account is able to make such a case only by virtually ignoring vast swaths of evidence from other fields, as well as his field's current troubles (which go unmentioned).

This priming-is-almost-everything tendency pops up everywhere in Bargh's book. In one section, he mentions a clinical trial involving sixteen severely depressed patients at a mental-health facility that appeared to show that some of their depressive symptoms improved after a session of "hyperthermia" treatment—that is, being exposed to very high temperatures. Bargh describes the study as "encouraging news" and then goes on to speculate about its potential applications in real-world settings given that not everyone can afford psychotherapy: "After all, it turns out that a warm bowl of chicken soup really is good for the soul, as the warmth of the soup helps replace the social warmth that may be missing from the person's life, as when we are lonely or homesick. These simple home remedies are unlikely to make big profits for the pharmaceutical and psychiatric industries, but if the goal is a broader and more general increase in public mental health, some research into their possible helpfulness could pay big dividends for individuals currently in distress, and for society as a whole."[29]

There's nothing wrong with studying hyperthermia treatments further, of course. But setting aside the comparison of sixteen individuals so unwell they are in a psychiatric institution with individuals with everyday depression, what stands out here is the sanding down of that condition—one that is horribly insidious, multifaceted, and sometimes resistant to treatment, which is why it wrecks the lives of so many millions of people around the world—to a mere lack of "social warmth," as potentially remedied by soup. And don't plenty of depressed people already drink hot coffee every day, to no apparent salutary effect?

Here and elsewhere in his book, Bargh overextrapolates from findings generated in circumscribed contexts. This is unfortunately common in social priming: researchers often make unwarranted leaps from statistically significant differences generated in contrived experimental settings to claims about the important influence of primes in real-world behavior.

In instances where researchers have a pretty good idea why people do what they do, Bargh ignores much of what is known in favor of priming-forward explanations. For example, he highlights research by Elke Weber and her colleagues which found that, as Bargh sums it up, "when the current weather is hot, public opinion holds that global warming is occurring, and when the current weather is cold, public opinion is less concerned about global warming as a general threat." He argues that "Our focus on the present dominates our judgments and reasoning, and we are unaware of the effects of our long-term and short-term past on what we are currently feeling and thinking."[30]

One problem with this is that the research Bargh cites doesn't quite say what he says it does. In 2014, *Nature Climate Change* published

Weber and her co-authors' clever paper, which was premised on asking study respondents for their thoughts on climate change in different ways and comparing the results.[31] Yes, the recent weather seemed to nudge people's views on global warming a little, but the authors of that paper found mixed-at-best evidence that priming played a role. What seems to account for the correlation is a reasoning process along the lines of "Well, we had a really cool summer, so it certainly doesn't feel like the planet is getting warmer to me." That's an inaccurate and unscientific way to answer the question, but it isn't necessarily driven by Barghian-style unconscious priming. It's basically the availability heuristic, which was mentioned earlier: people are using evidence that's more cognitively accessible but lower quality (the recent weather) rather than less accessible but higher quality (climatological data). The question of whether this is a conscious or unconscious process is, as the co-author Lisa Zaval put it in an email to me, "well outside our scope of work." And in fact when the authors tested the role of priming directly, using a scrambled-word task in one case and comparing the results when people were asked about "global warming" versus "climate change" in the other, they got a statistically significant result only one out of three times.[32]

More important, if you were attempting to understand, starting from scratch, why people do or do not believe in climate change, the recent weather would be very far down the list of the most important factors. Researchers like Yale's own Dan Kahan have done fascinating work which suggests that the extent to which we latch onto beliefs has a great deal to do with whether those beliefs are tied to an identity that is meaningful to us. That is, once people imbue a belief with a link to their identity—whether political or ethnic or religious—they will often jump through cognitive hoops to *ignore* experts who challenge it. In the United States, climate change is a deeply politicized subject (much to the planet's detriment). If you're a Democrat and/or a liberal, part of your social and political identity likely entails belief in the seriousness of climate change, whereas if you fly the GOP and/or conservative flag, part of your social and political identity likely entails some degree of skepticism about the concept itself or the severity of the threat it poses. In either case, defecting from the orthodox view of your "tribe" can have serious social consequences.[33] That's part of the reason why partisanship, Kahan has argued, is an overwhelmingly important force in determining what people say they believe and how they vote—a theory clearly borne out by polling. So at the level of survey research, yes, there is a statistically significant correlation between weather and concern with climate change. But even if unconscious priming could fully explain this phenomenon—and that's debatable, at best—there is little available evidence to think that this, on its own, will flip someone's vote, or cause them to rethink their political allegiances, or to do anything else that matters much in the real world. "Belief in climate change" as measured on a seven-point scale in public-opinion research is vastly different from "belief in climate change" as measured by real-life political behavior.

Bargh makes a similarly superficial argument about the primacy of priming in his discussion of the massive decline in crime in New York City (and the rest of the nation) that has shocked researchers and the public as it has unfolded since the 1990s; over that span, as we saw in chapter 2, the rates of murder and various other forms of violent crime have dropped dramatically. To explain it, Bargh turns to Rudy Giuliani and the extremely controversial "broken windows" policing protocol he championed, which involved aggressively policing minor infractions like public intoxication. Bargh explains that according to his interpretation of this policing practice, exposure to cleaner streets and fewer small-scale acts of mischief on the part of their fellow citizens could cause people to behave more lawfully. This, too, is a priming theory: Bargh puts it in the context of other work suggesting that subtle environmental cues can affect how people act, particularly with regard to behavior like littering. "There are of course other theories to explain this dramatic drop, and additional reasons for it," writes Bargh, "but it is also hard to argue with the positive consequences of a cleaner and more civil daily environment." Other research on the subject of priming and small-scale acts of disobedience, he argues, suggests "that the city's resurgence in turn was a result of a new culture of cues for positive behavior being instituted."[34]

Other than his brief, vague mention of "other theories," Bargh simply doesn't explore the many more serious ideas that researchers have come up with to help explain this crime drop; they range from the phasing of lead out of gasoline to changes in demographics to the dissolution of crack markets to other factors. To the extent that broken-windows policing was effective, researchers have argued that it worked by removing from the streets, via arrests for lower-level infractions, men who were either already wanted for more serious crimes, or who would have been likely to commit such crimes later.[35] But the claim that broken windows brought any direct crime-reduction benefit was seriously contested, anyway, in an important 2005 paper by Bernard E. Harcourt and Jens Ludwig, who argued that data from multiple cities offered "no support for [the] simple first-order disorder-crime relationship" that underpins broken-windows policing in the first place.[36] In any case, there is little support among serious criminologists for the idea that simply seeing unbroken windows and unlittered streets makes people less likely to commit the serious crimes, like murder and rape, that defined New York's scariest years. To the extent that social cues affect people's propensity for serious criminality, it is a much more complicated process than looking around and seeing whether streets are clean.[37]

It's also worth pointing out, as Bargh fails to do, that broken-windows policing had a terribly pernicious effect on some New Yorkers—particularly darker-skinned and poorer ones, who bore the brunt of the sudden enthusiasm for arresting people over low-level offenses. Broken windows also gave rise to "stop and frisk," the NYPD policy of searching individuals for no reason at all that became even more notorious among criminal justice reformers for the racially discriminatory manner in which it was administered. Even setting aside the moral problems with these styles of policing, they likely fostered significant distrust between police and various communities because of how often people were getting stopped, frisked, questioned, and sometimes arrested on thin or nonexistent pretenses.

This distrust could actually increase crime in the long run by reducing the likelihood distrustful residents of certain neighborhoods will call or cooperate with the police. In one important study built from real-world data, for example, a team led by the MacArthur-Grant-winning sociologist Matthew Desmond found that 911 calls in black neighborhoods of Minneapolis dipped after a highly publicized police assault of a black man.[38] In focusing on primes at the expense of everything else, Bargh ignores the broader context and ramifications of a very important, very controversial policy that no one who studies the matter closely views as having much to do with unconscious priming.

All of this is absent from *Before You Know It*. "Some situations induce us to be more polite and peaceful, others to be more rude and hostile," writes Bargh toward the end of this chapter. "Some imitative behaviors, such as dishonesty, can lead to financial meltdown, as with greedy investment bankers, while others can lead to the renaissance of a city, as when Mayor Giuliani and his fellow New Yorkers 'sweated the small stuff.'"[39] This is Primeworld: Big, potent structures are invisible. So is power. Policy analysis and trade-offs and second-order effects, too (except to the extent they involve priming). We're all just disconnected, free-floating individuals buffeted here and there by primes. Those bankers just needed more honesty primes, in much the same way inner-city kids caught in the crack trade needed more don't-murder-your-peers primes—in the form of graffiti-free walls.

Even if every single study Bargh mentions in his book were bulletproof, that wouldn't necessarily mean it would be fair to extrapolate from them that unconscious priming is as important in the real world as he says it is. Again, just because I can generate a statistically significant difference in a lab situation between those who are and aren't exposed to a given prime doesn't mean that priming is a particularly influential factor in their behavior, especially outside my lab. As the psychology researcher Tom Stafford put it in a 2014 article criticizing priming research, "Evidence of *differences* due to unconscious processes at the group level do nothing to confirm the importance of the unconscious processes in affecting the *overall response* of each individual."[40]

But the studies Bargh uses to build his case aren't bulletproof anyway. In fact, social priming—including many of the studies mentioned and conducted by Bargh himself—sits at the very center of psychology's replication crisis, which has called countless solid-seeming research results into question. Social priming is far from the only corner of psychology beset by replication travails, but in many ways it's a worst-case exemplar of what can happen when serious quality-control issues seep into one particular area. For that reason, it's impossible to fully understand how social priming and its many wild claims caught on without also understanding what's broken about modern psychological research.

MANY PSYCHOLOGISTS TRACE the start of the replication crisis to a 2011 study by the social psychologist Daryl Bem that claimed to offer evidence for so-called psi, or psychic power. Published in the leading *Journal of Personality and Social Psychology* and based on a decade's worth of experimentation, "Feeling the Future: Experimental Evidence for Anomalous Retroactive Influences on Cognition and Affect" reported that in eight out of nine experiments Bem had found evidence for "the anomalous retroactive influence of some future event on an individual's current responses."[41] In one study, for example, respondents had been able to predict, with surprising statistical accuracy, which of two computer images of a curtain was obscuring a pornographic image. Because the computer program didn't actually flip the metaphorical coin that determined the answer until after the subject input their guess, the result lent credence to the bizarre idea that the respondents had been able to predict the future.

This paper rocked social psychology. That was partly because many if not most of Bem's peers didn't believe what he found could possibly be true, and partly because the field was still reeling from a recent scandal that also involved too-good-to-be-true results. Diederik Stapel had been a wunderkind of Dutch social psychology who "made a name for himself with a series of eye-catching claims, many of them sitting firmly within the area of social priming: the presence of wine glasses improves table manners; messy environments promote discrimination; and, most recently, meat eaters are more antisocial than vegetarians," as Paul Jump put it in *Inside Higher Ed*.[42] But some of his young research assistants began noticing irregularities in his work, and by 2011 he had admitted to having fabricated data in his work. "The misconduct goes back to at least 2004 and involves the manipulation of data and complete fabrication of entire experiments," noted an article in the American Psychological Association's *Psychological Science Agenda*.[43] In the end, a shocking fifty-eight of Stapel's papers were retracted,[44] and the incident caused a great deal of soul-searching within the field. How had such rampant fraud gone on for so long? Perhaps, some argued, it had something to do with how cute and counterintuitive so many of Stapel's findings were—with just how irresistible to the public they were.

Bem didn't commit any sort of Stapelesque fraud, but statistically minded psychologists quickly poked holes in his paper.[45] As a result of the radical nature of its claim about ESP, as well as the responses from various skeptics, "Feeling the Future" came to exemplify what turned out to be a far more endemic problem in psychology than outright fraud: questionable research practices, or QRPs. QRPs refer to certain techniques for collecting, analyzing, and sharing data that can generate or lend credence to false-positive findings, and there is a direct connection between this issue and the question of which studies enjoy mainstream attention. The psychologist Chris Chambers's 2017 book, *The Seven Deadly Sins of Psychology: A Manifesto for Reforming the Culture of Scientific Practice*, is the best comprehensive explanation of the problem, and I highly recommend it, particularly for readers with an appetite for slightly more statistically complex accounts of psychology's internal problems.

One of the most prevalent and important QRPs is "hidden flexibility." "Faced with the career pressure to publish positive findings in the most prestigious and selective journals, it is now standard practice for researchers to analyze complex data in many different ways and report only the most interesting and statistically significant outcomes," writes Chambers. "Doing so deceives the audience into believing that such outcomes are credible, rather than existing within an ocean of unreported negative or inconclusive findings."[46] A simple, hypothetical example can make this more concrete: If I sell you a pill on the basis of data showing it reduces blood pressure relative to a control group given a placebo, but fail to tell you that I also tested its efficacy in improving twenty-five other health outcomes and came up empty on all of them, that's a very weak finding. Statistically, if you have enough data and run enough tests, you can always find *something* that is, by the standards of the statistical tests psychologists use, "significant." As Gregg Easterbrook put it in a quotation that Chambers includes in his book, "Torture numbers and they will confess to anything."

Or, to take a real-world example: Earlier in the chapter, I mentioned a study which found that "priming people with brief exposure to the brand Lipton Iced Tea—too quickly for it to be consciously noticeable—made them want that beverage more, albeit only when they were already thirsty and only for a short period." That last bit, the "albeit" phrase, is a potential red flag, an example of where hidden flexibility can rear its head. If the researchers found that there was no overall effect of exposure to Lipton Iced Tea on the respondents' desire for the beverage, they could then compare other, smaller *subgroups* (those who were younger or older or, say, already thirsty, or male, or whatever) with the control group to see if any of those comparisons bore fruit. Theoretically, they could slice and dice their data in this manner for a while until *something* popped out, because something will almost always pop out eventually. I'm not saying these particular researchers did that. The point is that unless researchers publicly document all the comparisons they made during what is often called their period of "exploratory analysis" of their data—which has not been the norm, traditionally—there's really no way to know whether their particular numbers have been tortured.

Another, related QRP is the practice of hypothesizing after the results are known, or HARKing. Let's say that going into an experiment, I hypothesize that priming children with images of Albert Einstein will cause them to perform better, relative to a control group of students not exposed to any primes, on a subsequent IQ test (because Einstein is associated with brilliance). Instead, I find the exact opposite: The Einstein-ed kids did *worse*, and by a statistically significant margin. Racking my brain in my tiny office, I eventually realize that actually what happened was that, because the photo was of Einstein as an old man, the children fell victim to that pesky elderly prime first discovered by John Bargh and his colleagues. It caused them to process information a little bit more slowly, as people do when they age. I write up my paper, complete with the "hypothesis" that Einstein priming would cause kids to perform worse on IQ tests, and without mentioning the original, failed hypothesis. It's published, and (inevitably) media outlets trumpet the finding that if kids are exposed to elderly primes, they'll do worse in school.

It should be fairly clear why this is problematic from the perspective of advancing science: my fellow researchers and the public alike are (1) denied access to the fact that my original theory failed and (2) misled into believing that the theory I switched to after the fact was what I had been investigating the whole time. By this standard, just about any finding can be explained, post facto, as though it were what the researchers expected all along.

By all indications, HARKing and other QRPs are quite prevalent in psychological science. In fact, the term likely caught on because of an important 2012 paper in *Psychological Science* in which a trio of researchers asked more than two thousand psychology researchers if they had engaged in QRPs: Sixty percent of them admitted to having intentionally excluded variables they had tested but that had come up nonsignificant, and half admitted to "selectively reporting studies that 'worked.'" Smaller but still disturbingly high percentages of the respondents admitted to other, even more serious QRPs.[47]

It's understandable why these practices caught on. If you want to succeed in psychology, you have to get published. If you want to get published, you have to report statistically significant findings. Contrariwise, you are unlikely to gain acclaim for publishing non-significant results (even though these results, too, provide valuable information). Chambers's book includes a fairly remarkable story about the Bem ESP study that demonstrates this nicely:

> Bem himself realized that his results defied explanation and stressed the need for independent researchers to replicate his findings. Yet doing so proved more challenging than you might imagine. One replication attempt by Chris French and Stuart Ritchie showed no evidence whatsoever of precognition but was rejected by the same journal that published Bem's paper. In this case the journal didn't even bother to peer review French and Ritchie's paper before rejecting it, explaining that it "does not publish replication studies, whether successful or unsuccessful." This decision may sound bizarre, but, as we will see, contempt for replication is common in psychology compared with more established sciences. The most prominent psychology journals selectively publish findings that they consider to be original, novel, neat, and above all positive. This publication bias, also known as the "file-drawer effect," means that studies that

> fail to show statistically significant effects, or that reproduce the work of others, have such low priority that they are effectively censored from the scientific record. They either end up in the file drawer or are never conducted in the first place.[48]

Humans respond to incentives. It's no wonder that QRPs became so commonplace.

BECAUSE OF QRPS, many studies that are technically "statistically significant" don't actually reflect real phenomena. Around the same time the psi study was published, the reform-minded psychology researchers Joseph Simmons, Leif Nelson, and Uri Simonsohn published a paper in *Psychological Science* in which they claimed to be able to *reduce people's ages* by having them listen to the Beatles' "When I'm Sixty-Four" rather than "Kalimba" (a song that "comes free with the Windows 7 operating system"), which the control group listened to and which was not associated with age-reversing properties.[49] Unlike Bem's study, this one was published not in earnest but to demonstrate how QRPs could lead to sexy but impossible results. "They achieved this result by analysing their data in many different ways, getting a statistically significant result in one of them by simple fluke, and then not reporting the other attempts," reported the British science journalist Tom Chivers in *Nature*. "Such practices, they said, were common in psychology and allowed researchers to find whatever they wanted, given some noisy data and small sample sizes."[50]

Suffice it to say that if you can "prove" a song can Bejamin Button experimental subjects using the standard statistical tools of psychology, you can "prove" just about anything. Simmons and his colleagues argued that, as a result of psychology's prevailing, QRP-friendly norms, things were truly dire. "In many cases, a researcher is more likely to falsely find evidence that an effect exists than to correctly find evidence that it does not," they wrote in their paper.

And perhaps because social priming had come to rely so much on a certain brand of sexy, counterintuitive finding, it turned out to be an epicenter of what came to be called the replication crisis. It was Daniel Kahneman who helped shine a harsh but much-needed spotlight on that corner of psychology. You'll recall that Kahneman lent his considerable imprimatur to remarkable social-priming findings with some very strong words in their favor. "Disbelief is not an option," he wrote in *Thinking, Fast and Slow*.[51] But the year after his book came out, in 2012, he wrote an open letter to a collection of social-priming researchers, including Bargh himself, which was subsequently published by *Nature*. "My reason for writing this letter is that I see a train wreck looming," he explained. Kahneman ticked off the various reasons social-priming results seemed particularly vulnerable at that moment, and explained that for those "reasons, right or wrong, your field is now the poster child for doubts about the integrity of psychological research. Your problem is not with the few people who have actively challenged the validity of some priming results. It is with the much larger population of colleagues who in the past accepted your surprising results as facts when they were published. These people have now attached a question mark to the field, and it is your responsibility to remove it."[52]

The easiest way to remove such a question mark is by replication. The theory is that if a psychological effect discovered in an experiment actually exists, rather than being the result of statistical noise, then it should show up in future, similarly conducted experiments. If, on the other hand, replication attempts repeatedly come up empty, that can bolster the case that the original effect was a meaningless statistical artifact, a thirsty traveler's mirage of a desert oasis. Thanks to Bem's study, the Stapel affair, and a more general ramping up of interest in methodological reform among psychologists, replications became increasingly common early in the second decade of the twenty-first century. More replications led to more internal and external discussion of replication, which led to more replications, and so on; Bem's study, shoddy as it was, helped spark a virtuous cycle.

Many priming studies have failed to replicate, including, in 2012, Bargh's elderly-walking result.[53] The statue study, too, failed to replicate. "In hindsight, our study was outright silly," one of the coauthors admitted to *Vox*, to his credit.[54] As did the study about money making people more capitalistic. And the one in which priming kids with the word "professor" increased their performance on a quiz.[55] And the one about flags making people more conservative.[56]

The Himmicanes study, sadly, turned out to be bogus, though that one didn't even need to be replicated. Not long after its publication, the Columbia University statistician Andrew Gelman, a frequent and gleeful debunker of shoddy science on his well-read blog, started poking holes in it, and by 2016 the researcher Gary Smith had published an article in *Weather and Climate Extremes* explaining, straightforwardly and compellingly, that the conclusion of the original study was invalid.[57] (As replications have grown, so too has the habit of simply taking a closer, more critical look at published findings to see where the authors might have erred. Recall that this sort of reanalysis, not a failed replication, called the seminal finding about screening during orchestra auditions into question.)

Some of the largest and most impressive replication attempts have been coordinated by the Center for Open Science in Charlottesville, Virginia, which is led by the University of Virginia psychology researcher and open-science trailblazer Brian Nosek. The first Many Labs replication project—which, as the name suggests, attempted to replicate past findings in a variety of different labs—was published in 2015.[58] During that effort, researchers attempted to replicate a hundred psychological studies that had been published in "three high-ranking psychology journals" in 2008. Their results, published in the top journal *Science*, revealed that they were able to successfully replicate the studies only a third to a half of the time—a very, very bad sign for the field.

The results of Many Labs 2, published in 2018, weren't much better: "Overall, 14 of the 28 findings failed to replicate despite the massive sample size, with more than 60 laboratories contributing samples from all over the world to test each finding."[59] (The details sometimes vary, but a study is generally considered to have replicated if it returns a statistically significant effect size pointing in the same direction as the original study that is approximately as strong as or stronger than the original finding. There is naturally some debate over what exactly the benchmark for a successful replication should be.)

It doesn't appear anyone has quantified the replicability of social-priming research compared with other types of research, but the outlook appears to be abysmal. Nosek is a careful, measured speaker when he talks with the media, and yet here's what he said to *Nature*'s Tom Chivers about social priming in 2019: "I don't know a replicable finding. It's not that there isn't one, but I can't name it."[60] In a May 2020 email Nosek confirmed he wasn't aware of any attempt to track the specific replicability of social priming. But for what it's worth, social psychology, the home base of social priming, fared particularly poorly in the original Many Labs effort: "Combining across journals, 14 of 55 (25%) of social psychology effects replicated," which was worse than the already-dire rate observed in other areas of psychology.[61]

In addition to actually replicating a given study, it's possible to estimate the *probable* replicability of that study based solely on its statistical parameters. To oversimplify: if a study shows a strong effect but has a tiny sample size, it's less likely to replicate. Ulrich Schimmack worked with Jerry Brunner to develop a tool called a Z-curve that explores this issue;[62] the user inputs the statistical parameters of a given published study, and out pops an estimate of its probable replicability. Schimmack took the impressive step of Z-curving every study in Bargh's book for what he called a "quantitative book review."[63] This, too, suggested troubles for social priming: Schimmack estimated the replicability of the studies Bargh cites to be about 41 percent. Now, the true value could be higher or lower. But it is reasonable, in light of both Schimmack's methodology and social psychology's empirically demonstrated replication travails, to expect that if someone attempted to replicate every study mentioned by Bargh, it is unlikely they would achieve coin-flip-level replicability.

None of this means that the concept of priming (as opposed to social priming) is bunk. There are a core set of results from the original study of priming that have held up over the years. Priming people with certain words, for example, does appear to affect their performance on subsequent, seemingly unrelated tasks, as laid out at the beginning of this chapter. That's a valuable insight into human cognition, as far as it goes, but one far removed from the idea of subtle influences having large and unconscious effects on real-world behavior.

The original Many Labs publication was viewed as a bombshell by many scientists in both psychology and other disciplines, but it was met with an assertive rejoinder published early the following year. That article, also published in *Science*, was co-authored by an impressive group of researchers, including the psychologist Dan Gilbert (who blurbed Bargh's book) and his Harvard colleague, the political scientist and research-methodology expert Gary King. They argued that the replicators had *themselves* made crucial methodological errors. Perhaps most important, many of the attempted replications differed, in important ways, from the original experiments. "An original study that asked Israelis to imagine the consequences of military service was replicated by asking Americans to imagine the consequences of a honeymoon; an original study that gave younger children the difficult task of locating targets on a large screen was replicated by giving older children the easier task of locating targets on a small screen; an original study that showed how a change in the wording of a charitable appeal sent by mail to Koreans could boost response rates was replicated by sending 771,408 e-mail messages to people all over the world (which produced a response rate of essentially zero in all conditions)," they wrote.[64]

While in other instances Many Labs replicated studies more

closely, it's certainly true that for various logistical reasons these and other alterations were made to the original designs. This raises important questions: How closely should a replication mirror the original finding? When does a failed replication offer strong evidence that the effect in question doesn't exist, and when does it mean something else—that the experiment wasn't conducted in exactly the same way, that it works only on certain samples, or that bad statistical luck interfered? There are no easy answers here, and the unresolved nature of these debates has helped fuel strong counterclaims from replication-crisis skeptics like Gilbert. He, and a handful of other established researchers, have suggested that there's no real replication crisis at all, or that its existence has been overblown by envious and opportunistic researchers seeking to tear down the work of others. "While academics wring their hands about a replication crisis, business, government, law and medicine are putting psychology's discoveries to work to improve the human condition," Gilbert told *Discover* magazine, offering what seems like a rather optimistic gloss on the situation given the hundreds of millions of dollars likely spent on the IAT and Comprehensive Soldier Fitness alone.[65]

For what it's worth, Many Labs has been generally open and responsive to the most substantive critiques of replication-crisis skeptics and has sought to investigate them empirically. Many Labs 2, for example, hewed much closer to the original study designs and was set up to evaluate whether using different samples during a replication attempt affected the likelihood of replication. That effort found that "if a finding was not replicated, it failed to replicate with little variation across samples and contexts," the project's authors wrote. "This evidence is inconsistent with a popular explanation that failures to replicate in psychology are likely due to changes in the sample between the original and replication study."[66]

On the whole, while the most ardent skeptics of the replication crisis have made certain fair points, some of them are trying to have it both ways. Psychology's biggest stars routinely claim, in successful books and TED Talks, that effects discovered in labs or small-sample-size real-world settings have clear, exciting implications for us all. And then, when those studies fail to replicate, some of those same researchers turn around and say *of course* those effects are quite fragile, easily undone by a different sample or different researchers or a different experimental protocol. But how can these effects matter in the real world if they can be snuffed out so easily by a gentle gust of methodological wind?

It would be one thing if the replication crisis had emerged at a time when psychological science seemed otherwise healthy, when the most popular lab ideas were regularly generating real-world interventions that were obviously effective. But that quite clearly isn't the case. Rather, a great many psychological findings burst onto the scene in a blaze of excitement only to sputter and stall out years later, and it's clear this is partially due to the prevalence of incentives to overclaim, *p*-hack, and engage in other acts that are detrimental to the entire project of behavioral science. As a result, it feels very risky to downplay the seriousness of the replication crisis, even if the science of replication itself is still growing up.

WHICH LEAVES US WHERE, EXACTLY? Should researchers stop publishing one-off studies? Should universities stop promoting them? Should scientists only apply findings that have been replicated many times? None of this seems practical or realistic. But there are some other, more tenable changes that can be made to the way psychological science is conducted that could reduce the probability faulty results are published in the future.

Since the replication crisis first started casting doubt on a broad range of psychological findings, some in the field have responded by calling for changes in the way data are collected, analyzed, and shared. To recycle an analogy, if replications are the equivalent of curing a disease that has already spread, patient by patient, these reforms are the equivalent of inoculating the population so as to forestall future outbreaks.

One key innovation is preregistration, or the practice of researchers publicly posting their hypotheses and data-analysis plans before getting to work. This means that if they switch their theory midstream from "Einstein priming will make kids smarter" to "Einstein priming will make kids less smart," or slice their data in questionable ways, these deviations from the publicly posted research plan will be apparent to all, disincentivizing HARKing and other questionable research practices. (The Center for Open Science has built a popular website that allows researchers to preregister their studies.[67]) Another is data sharing. Surprisingly, making one's data available to other researchers to look for errors and limitations in one's analysis has not, historically speaking, been a common practice in psychology. Until the recent past, it was even considered rude in some quarters: *You want my data? What are you accusing me of?!* To be fair, sometimes there are good reasons for keeping experimental data private, such as copyright/ownership issues or the arguably too-strict precautions demanded by institutional review boards to protect subjects' anonymity.[68] (IRBs are the bodies that give academic and government research an ethical green light before it is conducted.) Often, however, there aren't. The more norms shift toward data sharing, the harder it will be for questionable studies to be published, or for them to survive all that long without being debunked.

One of the most ambitious reform ideas is known as Registered Reports. It introduces an entirely new layer of peer review, earlier in the process, as this graphic from the Center for Open Science shows.

In this model, a study is accepted by a journal not on the basis of its *results* but on the basis of its *design*. That is, the journal commits to publish the study before knowing how it will turn out. If widely adopted, this would mark a potentially revolutionary inflection point. As Chris Chambers puts it in a quotation on the COS page, "Because the study is accepted in advance, the incentives for authors change from producing the most beautiful story to the most accurate one."[69]

There's another, more fundamental proposed reform to experimental psychology that deserves mentioning here: changing the standard statistical techniques researchers use to measure the likelihood that an observed effect is real. The "light" version of this is to simply bump the threshold for statistical significance from $p < 0.05$ to something lower, the idea being that this would go a long way toward addressing hidden flexibility issues: it may be easy to contrive $p < 0.05$ findings, but if you make the standard an order of magnitude tighter—$p < 0.005$ is one common suggestion—meeting it becomes significantly more difficult. (As you might recall, $p = 0.05$ simply means that if the so-called null hypothesis is true—there's no true effect being observed—there's a 5 percent or lower probability that a result at least as large as the one that was discovered would have popped up anyway. The 5 percent cutoff point is purely a matter of convention and tradition.)

The "heavier" version of this reform involves throwing out so-called significance testing altogether in favor of its longtime competitor: Bayesian inference. The difference between the two methods is explained quite nicely in a 2019 article in *Nautilus* by the mathematician Aubrey Clayton in which he critiques "the flawed reasoning behind the replication crisis."[70] Clayton writes of the aforementioned study involving *The Thinker*: "We need to assess the likelihood, before considering the data, that a brief encounter with art could have such an effect. Past experience should make us pretty skeptical, especially given the size of the claimed effect, about a 33 percent reduction in average belief in God." Were an effect like this as powerful as the numbers suggested, he points out, "we'd find any trip to a museum would send us careening between belief and non-belief. Or if somehow 'The Thinker' wielded a unique atheistic power, its unveiling in Paris in 1904 should have corresponded with a mass exodus from organized religion. Instead, we experience our own religious beliefs, and those of our society, as relatively stable through time." The same logic applies, of course, to many other incredible-seeming social-priming findings, like the idea that brief flag exposure would douse white-hot political divisions among Israelis.

Traditional statistical testing (known as Fisherian significance testing, after its founder, Sir Ronald Fisher) does not take into account one's prior belief about the likelihood of a given result. Bayesian inference—named for its founder, the eighteenth-century English cleric-*cum*-statistician Thomas Bayes—does. The general idea is deeply intuitive: the more improbable a result, the more evidence you need before you should accept it.

I'll relegate the actual equation underlying this idea to an endnote,[71] but the key point here is as important as it is simple: if you use a given piece of evidence to try to determine the likelihood X is true without factoring in the probability you would have assigned to X being true *before* encountering that evidence, you could be led astray. The most classic example of this is in medical testing: Imagine you get a positive test for a medical condition. One interpretation is that you have the condition, because the test says you do and it has, say, only a 5 percent false-positive rate. Another interpretation—the more Bayesian approach—is that you *might* have the condition but that you need to read the test result not only in light of the possibility of a false positive but also in the full context of the condition's *overall rarity*. Whether the disease in question infects one out of every hundred Americans or one out of every million matters a great deal to establishing the likelihood that your positive result means what it appears to mean. In some cases, depending on the specific numbers, a false positive may be more likely than your having the condition—a rather important fact that could be obscured entirely in the absence of Bayesian reasoning.[72] Similarly, in psychology, there are situations in which a "positive" result might still leave a researcher with a very high probability of having found nothing at all, simply because we have good prior reasons to believe statues are unlikely to drive churchgoing folk into the arms of Richard Dawkins.

We already know that thanks to questionable research practices, even weak data can be fairly easily wrestled down below the $p < 0.05$ threshold. When a hypothesis is instead evaluated via Bayesian inference, this has the effect of adding a check on reckless or sloppy experimentation and data analysis. Clayton points out that in the statue study, simply switching from Fisherian significance testing to Bayesian inference, using otherwise the *exact same data*, would have led to a very different outcome—one that would have saved everyone a lot of time. According to traditional significance testing, he explains, there was only a 3 percent chance the difference between the groups could be explained by chance alone. But if you account for a prior estimate of the probability that statue viewing can sap one's faith in God, suddenly the result swerves wildly in the other direction. This is true even if one is fairly forgiving in making what is an admittedly subjective assessment about the likelihood of statue magic. "Maybe we're not so dogmatic as to rule out 'The Thinker' hypothesis altogether, but a prior probability of 1 in 1,000, somewhere between the chance of being dealt a full house and four-of-a-kind in a poker hand, could be around the right order of magnitude," he suggests. If you run that prior probability and the researchers' other numbers through Bayes's theorem, you'll "end up saying the probability for 'The Thinker'–atheism effect based on this experiment was 0.012, or about 1 in 83, a mildly interesting blip but almost certainly not worth publishing."[73] (Likely because they anticipated controversy, the editors of the *Journal of Personality and Social Psychology* published a response to Daryl Bem's psi paper written by the Dutch psychologist

E. J. Wagenmakers and his colleagues in the very same issue which argued forcefully that taking a Bayesian approach to Bem's work showed that the "evidence for psi [was] weak to nonexistent."[74])

One potential weakness of the Bayesian approach is that there is an inherent subjectivity to estimating many sorts of "prior possibilities"; it's easy to imagine situations in which researchers disagree wildly about what an "appropriate" prior probability is. But no statistical tool is perfect, all involve arbitrary judgments of one sort or another (such as using the value 0.05 as an important benchmark), and the advocates for Bayesian inference insist that it will solve far more problems than it will cause. Plus, you could always report a range of estimates based on different assumed prior probabilities.

Now, it's important to note that just because a study result defies common sense is no reason to assume it's false; that would be the wrong lesson to draw from the replication crisis and some researchers' advocacy for a shift toward Bayesian inference. Throughout history, scientists have flouted the "common sense" of their times, sometimes sparking scientific revolutions and forging new kinds of common sense in the process. Take, for example, the concept of so-called quantum entanglement between two particles—a phenomenon in which "measuring one member of an entangled pair of particles seems to instantaneously change the state of its counterpart, even if that counterpart particle is on the other side of the galaxy," as one *Scientific American* article puts it. This idea—not to mention so much else from quantum mechanics—laughs in the face of the human conception of "common sense."[75]

But at the moment, it appears psychology has a much bigger problem with false-positive than with false-negative errors, and, therefore, that's where the focus of reform efforts should lie. Maybe a one-in-eighty-three chance will point to a real effect. To explore the possibility, researchers can always attempt to replicate the experiment, possibly with a new, more forgiving prior-probability estimate (because they now possess a smidgen of evidence nudging them toward belief in the statue effect). A true effect will, over time and repeated experimentation, win out, even if it defies common sense.

BRIAN NOSEK, THE CHAMPION of replication and transparency in psychological science, takes a positive view of psychology's future. "The credibility revolution in psychology has made enormous progress this decade and still has a long way to go," he wrote to me in an email. "A substantial portion of journals have adopted progressive transparency policies and innovations to improve rigor and reduce publication bias, such as Registered Reports. Funders are supporting metaresearch about the process of doing research and evaluating reforms. Most critically, the grassroots efforts of thousands of individual researchers—mostly early-career researchers—has irrevocably altered the norms and accelerated adoption of behaviors like preregistration and data sharing. Thanks to them, psychological science is in a different place today than it was in 2011. Psychology in 2031 is going to be amazing."

Nosek's point about younger psychologists seems particularly important. Norms matter a great deal, and an entire generation of psychologists, some of them already past their PhD years and embarking on their careers as young professors, has come of academic age during an era of increased concern about the integrity of psychological science. They are much more likely to know what *p*-hacking is than their predecessors, and much less likely to do it.

We'll have to wait and see whether Nosek's prediction is borne out. But things certainly seem to be trending the right way. One particularly strong data point is the current state of social priming itself: it appears that there has been a significant diminution of interest in the field. While it still has its defenders, "so many findings in social priming have been disputed that some say the field is close to being entirely discredited," reported Tom Chivers in his *Nature* piece.[76] Other than some ongoing replication attempts, there's been a clear shifting of attention and resources away from social priming and toward other areas. This is good: it shows that the psychological establishment is responding to the available evidence—weak, wobbly effects that can't be replicated—and acting appropriately. These days, if someone publishes a study claiming that statues can turn people into atheists, or that people ignore warnings of feminine-sounding hurricanes, they will have significantly more trouble getting glowing media write-ups, and their peers will be a bit less shy about expressing their public skepticism. All of which are signs of progress.

# 8

# NUDGING AHEAD

In 2014, social scientists and criminal justice reformers were trying to figure out how to solve a particularly frustrating problem with New York City's justice system: the surprising frequency with which those summoned to appear in court failed to do so.

For a long time, under New York City law, certain low-level offenses were handled at the discretion of the officer who discovered them. If you were caught by a police officer drinking or urinating in public, or committing other minor infractions, that officer could arrest you on the spot. Alternatively, he or she could fill out a so-called criminal summons and hand you a carbon copy of it: you were now summoned to appear in court at an appointed time and enter your plea of guilty or not guilty. The only exception to this requirement was for public drinking or urination, in which case you could indicate on the form that you pled guilty and agree to a fine of $25 for drinking or $50 for urination and then mail in the form with a check. (In March 2016, the Manhattan district attorney and NYPD announced that they would switch to a summons-only system in that borough for certain minor offenses, including public drinking, avoiding arresting people except when there was a "demonstrated public safety reason to do so."[1])

These laws were eventually reformed, but over the years many New Yorkers had warrants issued for their arrest for failing to appear in court to contest $25 fines. Suffice it to say that, particularly for the many residents of my city who are poor, precariously employed, or both, a bench warrant can cause a life-upending disaster; people often lose their jobs and take severe financial hits from the snowballing difficulties that getting arrested can bring with it.

Given these stakes, it was genuinely shocking how many people failed to show up at court or pay their fines. In 2014, the rate of nonresponse was 41 percent. That's why the city government asked ideas42, a behavioral science nonprofit, and the University of Chicago Crime Lab "to design and implement inexpensive, scalable solutions to reduce the failure to appear (FTA) rate," as those organizations would later explain in a co-authored report.[2]

The researchers quickly homed in on a likely culprit: the design of the summons form itself. You can see the original form on ideas42's website, and if you do, you will likely agree that it is a bit difficult to parse and understand.[3] It doesn't do a particularly good job imparting the single most important piece of information it contains: that you, the recipient, *must* show up at court (unless you have been accused of public urination or drinking and plead guilty by mail).

So the team got to work redesigning it. In the new version, which you can also view online, certain elements pop off the page in a way they didn't before. The form's title was switched from the vague

"Complaint/Information" to the much more specific "Criminal Court Appearance Ticket." The most crucial information was moved from the back of the form to the front, right near the top, and other tweaks were made as well.

The researchers found that the form redesign brought the FTA rate down from 41 percent to 36 percent. This might not sound impressive but as they note, it's the equivalent of seventeen thousand fewer arrest warrants per year. Among the subset of summoned individuals who provided their cell phone numbers to the officer who cited them, the researchers also experimented with sending different types of text messages reminding them to show up at court. The most effective text message, on its own, reduced the FTA rate to 26 percent.[4]

Once these changes were made, they required little further cost: the form redesign was a onetime expense, and the text-messaging system could be almost entirely automated.[5] The bang for the buck here was quite impressive. Because of these relatively modest changes, the researchers predicted, over the years hundreds of thousands of fewer people would be at risk of arrest in New York City, helping to ameliorate the city's considerable court backlog.

This is the appeal of what behavioral scientists call nudging.

THE NEW YORK FORM-REDESIGN CASE highlights one of the most surprising aspects of human nature: just how easily we are waylaid on the path to good decision-making. As the authors of the Chicago Crime Lab/ideas42 report put it, attempting to explain New York's high failure-to-appear rate, "People forget, they have mistaken beliefs about how often other people skip court, they see a mismatch between minor offenses and the obligation to appear in court, and they overweigh the immediate hassles of attending court and ignore the downstream consequences."

If people were more rational, such behavior would be rare: most people can afford a $25 or $50 fine, and if asked via survey whether they would be willing to pay that sum to avoid arrest, it's undoubtedly true that almost everyone would answer yes in an instant. But the world distracts us, and our brains sometimes perform poorly at processing information and making decisions in an adaptive manner.

Nudging arose because researchers and policy makers started to better understand these facets of human nature, as well as the ways "choice architecture"—the manner in which choices are presented to people—could be altered to facilitate better, or more intentional, decision-making. It is the brainchild of the Harvard University legal scholar Cass Sunstein and the University of Chicago economist Richard Thaler. In their 2008 mega-bestseller, *Nudge: Improving Decisions About Health, Wealth, and Happiness*, they define a nudge as "any aspect of the choice architecture that alters people's behavior in a predictable way without forbidding any options or significantly changing their economic incentives. To count as a mere nudge, the intervention must be easy and cheap to avoid. Nudges are not mandates. Putting fruit at eye level counts as a nudge. Banning junk food does not."[6]

Other common and reasonably well-studied nudges include sending people text-message reminders to take certain actions (vote, study, show up at court, and so on), providing them with information that will make them feel like a social outlier if they make a certain decision (say, notices on utility bills that read, "You use more energy than 72% of your neighbors"), or switching certain decisions, like whether to be an organ donor or to enroll in a company 401(k) plan, to opt out rather than opt in, meaning the default option on the relevant form is yes and one must make a specific effort to indicate that no, one does not wish to enroll.

Nudging is designed to harvest low-hanging fruit, to improve people's decision-making without introducing onerous or politically challenging laws or regulations. Many fans of nudging believe that when authorities determine from on high what isn't allowed, they often do more harm than good. But if you just move the desserts to the back of the cafeteria, rather than outlaw them altogether, that's a less fraught approach and one that still allows people to retain the ability to make decisions many would consider ill-advised if they're intent on doing so.

In a short period, nudging has become nearly a household term. Governments and other institutions all over the world have adopted nudge principles in a manner designed to maximize cost-effectiveness and to help people make healthier or otherwise wiser decisions. Barack Obama, famously, was one of the first heads of state to truly embrace this twenty-first-century behavioral science revolution; he applied nudge insights to various pieces of legislation, gave Sunstein a crucial role overseeing government regulation, and in 2014 set up a "nudge unit" in the White House. By some measures the British government was even quicker to adopt nudging; Prime Minister David Cameron formed what *The New York Times* called a "Ministry of Nudges" in 2010.[7]

More than many other concepts in this book, nudging is an idea with a fair bit of genuine empirical heft behind it—in part, as we'll see, because it piggybacks on some of the most important behavioral science of the twentieth century. Further adding to the appeal of nudges is the fact that because they tend to target specific behaviors in a narrow, well-defined way, it is generally straightforward to test their effectiveness.

In short, there are legitimate reasons for all the excitement over nudging. But a close look at this concept also reveals the limits of even a solidly founded behavioral science. There is only so much nudging can do, and if we put too much faith in its low-cost, low-effort interventions, we may miss what matters most.

MANY OF THE BEST and more theoretically well-founded nudges come from behavioral economics. In fact, nudging owes so much to this field that it probably wouldn't exist without it.

Some sources say that behavioral economics was founded by the economist Richard Thaler,[8] who won a Nobel Prize in 2017 for his discoveries.[9] But while his contributions are exceptionally important (and we'll return to them), this line of research and theory really began with Daniel Kahneman and Amos Tversky, two brilliant Israeli psychologists who met in the 1960s, formed one of the most productive partnerships in the history of science, and who would later mentor Thaler.

Behavioral economics was born from the gap between how humans really are and how social scientists long saw them. Mid-century social scientists concerned with human judgment and decision-making—particularly economists—generally relied on models that treated humans as fundamentally rational. Members of "homo economicus," or "econs" for short, have stable, clearly ordered preferences, aren't distracted by extraneous information, and behave, generally speaking, like machines programmed with elegantly designed, well-being-maximizing algorithms. "If you look at economics textbooks," write Sunstein and Thaler in *Nudge*, "you will learn that homo economicus can think like Albert Einstein, store as much memory as IBM's Big Blue, and exercise the willpower of Mahatma Gandhi."[10]

When Kahneman and Tversky were first establishing themselves as young researchers in Israel, social scientists understood that homo economicus was, at best, only a roughly accurate model, because research had shown that real-life humans simply weren't as good at certain types of reasoning as econs were theorized to be. The problem, back then, was that these scientists didn't particularly care about exactly what went wrong when people made irrational-seeming choices or judgments; it wasn't considered all that interesting a problem. Rather, the prevailing assumption was simply that people were, if not rational, rational-*ish*: that they did intuitively know how to follow the right statistical rules to maximize their profit or make other decisions correctly, but that they simply lacked expertise in doing so, almost like a child who knows all the individual mechanics of bike riding but can't stay upright for more than a block or two.

Kahneman first encountered this view when Tversky gave a visiting talk about it to his class in Israel, and he was instantly skeptical.[11] As Michael Lewis writes in *The Undoing Project: A Friendship That Changed Our Minds*, his history of the origins of behavioral economics, "In Danny's view, people . . . were not statisticians of any kind. They often leapt from little information to big conclusions. The theory of the mind as some kind of statistician was of course just a metaphor. But the metaphor, to Danny, felt wrong."[12] Tversky was quickly converted to Kahneman's view—that whatever was going on here was more interesting than the prevailing model of people as novice statisticians, and more worthy of systematic study—and off the two brilliant researchers went. They proceeded mostly by simply asking people, in studies, to make various sorts of judgments, the goal being to see whether the inevitable errors that popped out followed any particular patterns.

Kahneman and Tversky asked mathematically inclined psychologists to estimate a group's average IQ based on the score of one outlier; they asked college students which of two distributions of marbles was more likely in a hypothetical game in which marbles are distributed at random; they asked people whether a large or small hospital was likely to have more days in which 60 percent of the births were boys, or if the likelihood was roughly the same in each case.[13] They asked and they asked and they asked, comparing the answers people gave with the statistically correct ones.

This ended up being a legendary partnership—one that generated decades' worth of vital insights about human nature and that culminated in a 2002 Nobel Prize in economics for Kahneman, even though he is not an economist. Tversky surely would have shared the prize with him had he not died of cancer in 1996[14] (the Nobel committee does not award the prize posthumously, but did mention him in its announcement about Kahneman[15]). The pair showed, convincingly, that most humans do *not* have the right general idea about statistically informed decision-making, minus some random error. Rather, many of our errors appear to be predictable. For example, most people falsely believed that a large hospital and a small one had about the same likelihood of a boy-heavy day, but in fact it's the smaller one, because fewer births means a smaller sample size and a higher chance of statistically unlikely events. And even those quantitatively talented mathematicians consistently underestimated the impact of a single outlier on the rest of the group's average IQ.

Kahneman and Tversky were able to discover certain predictable mental shortcuts, or heuristics, people use to understand the world around them, often causing them to err. The *representativeness heuristic*, for example, is the tendency for people to rely on stereotypical information rather than anything like genuine statistical reasoning to gauge an event's probability. Take one famous study in which participants were asked to rank the likelihood a fictional person had certain characteristics. Study respondents were told about "Linda," who was "31 years old, single, outspoken and very bright," and who as a student "was deeply concerned with issues of discrimination and social justice, and also participated in anti-nuclear demonstrations." After they were provided with a list of possible descriptions of this woman, "the great majority of subjects," as Kahneman and Tversky put it, rated her more likely to be "a bank teller [who] is active in the feminist movement" than "a bank teller."[16]

Probabilistically, this is impossible: it can't be less likely for someone to be X than for them to be X and also Y. Here and in so many other contexts, Kahneman and Tversky's experiments demonstrated how people's minds could get snared on shiny but fundamentally irrelevant pieces of information: *Linda is outspoken and interested in social justice, which sounds like the sort of person who would be a feminist.* Econs would never get so easily waylaid by distracting informational baubles, but humans are different.

The *availability heuristic*, which we've already seen in action, is another vital insight from behavioral economics: all else being equal, the more cognitively accessible an event is, the greater people will rate its likelihood, which can, for obvious reasons, lead to extremely distorted perceptions of probability (children very rarely get trapped at the bottom of wells, but when they do, it's big news that sticks in many people's minds). If people were, in fact, generally competent at statistical reasoning, setting aside some random errors, something as statistically irrelevant as an event's cognitive accessibility would not sway their beliefs as significantly as it often appears to.

So-called framing effects, too, vividly demonstrate the potential real-world relevance of behavioral economics. Kahneman and Tversky first demonstrated this with their so-called Asian disease task. Study participants were given the following prompt: "Imagine that the U.S. is preparing for the outbreak of an unusual Asian disease, which is expected to kill 600 people. Two alternative programs to combat the disease have been proposed."

Here the experiment branched off into one of two different conditions. In the first, participants chose between Programs A and B: "If Program A is adopted, 200 people will be saved. If Program B is adopted, there is ⅓ probability that 600 people will be saved, and ⅔ probability that no people will be saved. Which of the two programs would you favor?" Other participants were asked to choose between Programs C and D: "If Program C is adopted 400 people will die. If Program D is adopted there is ⅓ probability that nobody will die, and ⅔ probability that 600 people will die."[17]

If you read the options closely, you'll see that Programs A and C are identical (in both cases, two hundred people live and four hundred die), and that programs B and D are as well (a one-third probability no one dies, and a two-thirds probability everybody dies). So logically, respondents' preferences should be the same in the two conditions. But that isn't what Tversky and Kahneman found: rather, people overwhelmingly chose A over B, and D over C. Whether the problem was framed in terms of gains (lives saved) or losses (lives lost) had a large impact on which option they chose—part of the broader phenomenon of *loss aversion*, or the human tendency to feel more keenly the threat of losing something than the prospect of gaining something.

It would take an entire book to lay out the key principles Kahneman and Tversky uncovered in a comprehensive fashion, and in fact such books have been written—*The Undoing Project*, *Thinking, Fast and Slow* by Kahneman himself, and *Misbehaving: The Making of Behavioral Economics* by Richard Thaler among them. The most important take-home message from this body of research is that people are, as the psychologist Dan Ariely would later put it in a book title, predictably irrational. There is a shape to the biased way they process information, and in some situations subtle tweaks to how that information is presented can spur changes in how it is processed, which can in turn spark corresponding changes to the thinking or behavior that ensues. (One could argue that homo economicus is, on the other hand, thought to be "unpredictably rational"; that is, he thinks in a straightforwardly robotic manner but makes random errors that cancel each other out along the way.)

Nudges, then, can be developed simply by understanding the difference between humans and econs. And if your reaction to some nudges is that they seem obvious, that's sort of the point; much of our choice architecture is designed with a faulty understanding of human decision-making, leaving an opening for nudgers to make tweaks that can, in retrospect, appear trivial. Sometimes, people build systems that econs would navigate just fine but that trip up humans unnecessarily. Like New York City's old summons system, for example: "Traditionally, criminal justice policy is informed by the assumption that people make an explicit decision to offend, and so most approaches aim to make crime less worthwhile," write the authors of the ideas42 and Chicago Crime Lab report. "But our interventions are built on the view that people who miss their court date do not necessarily make an active choice to skip it. Rather, they may have failed to consider the decision at all due to a number of obstacles."[18]

So that's the key insight of nudging; "obstacles" to making a good choice can be rendered less imposing via choice architecture. And usually for cheap.

LIKE ANY OTHER BIG, important innovation that threatens an entrenched status quo, Kahneman and Tversky's insights were adopted by different groups at different speeds. These concepts had a particularly difficult time making inroads among economists, who have, for a long time, been the most influential social scientists around; when you think of a close adviser to a prime minister, you probably think of an economist rather than a psychologist or a sociologist (even in this age of nudging). Economists and their equations long relied on homo economicus, and some economists clung to it well after it became clear that there were fundamental flaws in the model. (One could say they were falling for the sunk cost fallacy, another condition to which humans, but not econs, are susceptible.)

As Cass Sunstein notes in his memoir of government service, *Simpler,* when he started his academic career at the University of Chicago in the 1980s, the titans in the economics department and law school, such as George Stigler, Ronald Coase, Gary Becker, Richard Posner, and Frank Easterbrook, were extremely skeptical of this sort of thinking. "If you challenged the view that human beings are rational, in the economic sense, you would get a withering look from one or another giant. The look said, very simply, that anyone who dared to question the rationality assumption was, well, not rational."[19] These were important gatekeepers, because, like the denizens of other top-notch academic institutions, they often enjoyed direct access to the halls of power (the University of Chicago had, and has, one of the most famous economics departments in the world).

Eventually, though, behavioral economics broke through. Perhaps those economists just needed to hear the gospel from one of their own. "In the mid-1980s the cavalry arrived," writes Sunstein. "Its young commander was Richard Thaler, an economist at Cornell, who was writing some brilliant papers about what people are really like, and about how they depart from economic understandings of rationality."[20] Thaler became a protégé of Kahneman and Tversky, and his research helped transform economics—particularly his findings on "the tendency to not behave completely rationally, notions of fairness and reasonableness, and lack of self-control," as the Nobel committee summed up his most important work.[21] After he joined the faculty at Chicago in 1995, he and Sunstein formed a lasting, fruitful friendship and academic partnership.

These days, certain behavioral economics ideas are almost universally accepted among the most influential social scientists, and seemingly everyone is interested in nudging. *Nudge* itself had a massive impact. In fact, the UK's nudge unit was founded in part because David Cameron happened to pick up, and promptly fall in love with, a copy of the book. Barack Obama even made a point of self-nudging by altering his own choice architecture. In 2012, he famously told Michael Lewis that he had culled his wardrobe to just two colors of suits. "I'm trying to pare down decisions," he said, which was in context a clear reference to the theory that if we have too many choices, "decision fatigue" can set in and worsen our cognitive performance.[22]

Early in his presidency, Obama attempted to apply the insights of behavioral economics at a crucial moment. His 2009 stimulus bill was shaped by a key behavioral-economic principle first discovered by Thaler: *mental accounting*, or people's tendency to treat money differently depending on what mental "bucket" they put it into. All else being equal, when people consider money "wealth," they're more likely to save it, and when they consider it "income," they're more likely to spend it. (Mental accounting is another broadside against homo economicus, for whom $5 is $5 regardless of its provenance;

under the assumptions of that model, our brains treat money as fungible.)

The goal was for the stimulus money to be spent by its recipients, stimulating the economy, not socked away. "If you want people to spend the money, you don't want to give them one big check, because that makes it more likely that they'll think of it as an increase in their wealth and save it," explained James Surowiecki in *The New Yorker.* "Instead, you want to give them small amounts over time."[23] Which is what the administration did: On average, Americans who were keeping a close eye on their finances would have noticed that about $40 less per month was withheld from their income for tax purposes. The idea was that delivering the funds in this manner would make it more likely the money would fall into their *income* rather than their *wealth* bucket.

Not long before the stimulus bill passed, Obama picked Sunstein to head the Office of Information and Regulatory Affairs, which oversees all federal regulations and ensures they are cost-effective.[24] Nine months later Sunstein was confirmed, and he served as head of OIRA for three years, bringing with him an ardent enthusiasm for nudging and behavioral-science thinking as well (though in many circles he is just as well known for his advocacy for cost-benefit analysis).

In his second term, Obama sought to follow the 2010 lead of Downing Street's "Ministry of Nudges" and more formally institutionalize the practice of nudging within the federal government. In 2014, the White House established the so-called Social and Behavioral Sciences Team. The next year, Obama issued an executive order solidifying the organization's role within the government and directing federal agencies to apply its insights.[25]

In the first annual report of the SBST, described in a White House press release as "a group of experts in applied behavioral science that translates findings and methods from the social and behavioral sciences into improvements in Federal policies and programs for the benefit of the American people,"[26] its authors lay out a clear case not only for the team's existence but for the more general importance of applying sound psychological principles to governance. "Because SBST projects are designed to address only the behavioral barriers that affect how people engage with programs, project effects can be modest," the authors write. "Yet, because behavioral changes to program administration often require little or no additional cost, returns on investment can be large even when project effects are small. It is no more expensive to send an effective version of an email than an ineffective one."[27] And so we return, once again, to that idea of low-hanging fruit: When the government is already performing a given function, regardless of what year it is or who is in power, what's the argument against making it a bit more behaviorally informed, especially when, in the hands of a deft choice architect, it's so easy to establish a positive cost-benefit ratio?

The SBST's two annual reports explain the many pilot experiments the team ran. Some are so simple that it's shocking, in retrospect, that it required the creation of a team with a fancy-sounding name to implement them; in one experiment, for example, when certain government employees started a single-sided printing job, a dialogue box would pop up asking them to change their settings. "This prompt increased the likelihood of double-sided printing by 5.8 percentage points, from a baseline of 46 percent." Over time, that's a small tweak that saves a lot of paper, simply by nudging people to change paper-use habits, and giving them an extremely easy means of doing so.

The SBST team also sent emails to more than eight hundred thousand student-loan borrowers who had fallen behind on their payments. The emails clearly explained the option to enroll in so-called income-driven repayment plans, in which debts can be repaid at a slower rate, pegged to one's income, and led to a surge of applications to these plans. Overall, thirteen of the fifteen initiatives the SBST attempted in its first year were deemed successful, one unsuccessful, and one as having delivered ambiguous results.[28] Even when the overall effects of the interventions were small, so too were the costs—one of the primary appeals of nudging. The next year's annual report offered a similarly rosy assessment of the team's progress. Unfortunately, that annual report was the last, because after the Democrats lost the White House in the 2016 election, the SBST was disbanded by the Trump administration.[29]

In the UK, things have gone a bit differently; there, behavioral economics and nudging have managed to establish what appears to be a permanent beachhead in government. Part of this is due to the UK nudge unit's impressive launch. If the team "did not save the government at least ten times its running cost (£500,000 a year), it was to be shut down after two years," noted *The Economist* in 2017.[30] But the office ended up "saving about 20 times its running cost" and soon inspired imitators not just in the U.S. but around the globe as well; by 2014, the article notes, more than fifty countries had nudge-unit-like government bodies. In 2014, the Behavioural Insights Team spun off as its own company, still partially owned by the British public. It continues to consult for the British government but also has seven offices around the world.

BECAUSE A NUDGE CAN BE almost anything, it's difficult to make any declarative statements about the success of "nudging" itself. What is clear is that a broad set of nudges based on sound behavioral science principles have been fairly successful, at least in the sense of saving more money than they cost to implement and delivering positive, measurable changes to behavior.

Perhaps the clearest example is the case of employee savings programs like 401(k) accounts, particularly the practice of automatically enrolling employees and forcing them to opt out if they don't want to participate. Because of the cumulative effects of making small contributions to an account that benefits from compound interest, some of these nudges are associated with serious savings increases. This insight was eventually applied at a very high level: the 2006 Pension Protection Act required employers to automatically enroll employees in 401(k) plans.[31] According to one 2018 Vanguard study of more than five hundred thousand recently hired employees that compared their participation in various employee savings plans with and without automatic enrollment, "Automatic enrollment more than doubles participation rates among new hires. Over the entire period of our study, the participation rate for new hires was 91% under automatic enrollment versus 42% under voluntary enrollment."[32]

Another large-scale, real-world nudge effort involved Facebook. In 2010, the social-media giant examined the behavior of "61 million Facebook users during [that year's] congressional elections," as a subsequently published paper on the effort put it.[33] Users who were exposed to a notice indicating the names and profile photos of friends who said they had voted (by clicking an "I Voted" button some users experienced) were 0.39 percent more likely to vote, as validated by voter records, than those in both a control group and a group that received an informational notice about it being Election Day. That's a small-seeming percentage, but in an experiment of this size, it represents a potentially meaningful number of votes, especially in an off-year election.

Certain healthy-eating nudges have shown effectiveness, too. A 2019 meta-analysis divided the studies it examined into nudges that were "cognitively oriented," meaning they imparted information like calorie counts; "affectively oriented," meaning they tried to make people feel happier or more excited about making healthier choices (by organizing fruit in a pyramid, for example); and "behaviorally oriented," meaning they targeted behavior most directly (by making desserts a bit less accessible, for example). All three categories had statistically significant effects, but the behaviorally oriented interventions were particularly effective; study participants exposed to these nudges consumed 209 fewer calories per day compared with those in control groups.[34]

And some rigorous, broader scientific evaluations of behavioral economics principles have been encouraging as well: Assessing the "Asian disease task," one Many Labs replication effort, put the overall effect size of positive versus negative framing effects as a nudge tactic at somewhere in the neighborhood of $d = 0.5$ to $d = 0.6$, or "medium," to oversimplify. Many Labs found even larger effects when it attempted to replicate studies involving "anchoring," the tendency of people's numerical guesses to be affected by irrelevant information (for example, all else being equal, if you're exposed to a very high number before being asked to estimate the circumference of Earth, you'll provide a higher estimate than if you had been exposed to a very low number, even though such information is extraneous).[35] This is considered impressive by the standards of psychologists working on these sorts of interventions, which makes sense when you imagine the impact tweaking a question's wording could have when a survey or form will be filled out by many people. That is, if you multiply a medium-size effect delivered at low cost over many people, an intervention can yield impressive results.

But there have been plenty of nudge failures too, and it's important to note them, if only to keep in mind that just because a real-world intervention is based on a solid theory of human behavior, that doesn't necessarily mean it will be successful.

There was a lot of early excitement over the possibility of using nudges to increase the number of organs available for donation. In *Nudge*, Sunstein and Thaler have an entire chapter called "How to Increase Organ Donations," which argues for making organ donation automatic upon death and permitting people to opt out if they wish. Some countries went ahead and shifted their laws in this direction in the intervening years. But the results, overall, have been disappointing. One study published in 2019 compared organ donation rates in opt-in versus opt-out countries and found that "there was no significant difference in total deceased-donor rates in per million populations but there was a reduction in living donor numbers in the opt-out countries." That latter finding, argued the researchers, "suggest[s] that a simplistic switch to the 'opt-out' model has unintended consequences for living organ donation that does not provide a 'quick fix' to improve donor rates that has been previously suggested."[36] It appears that when people realize they have been defaulted into the status of organ donors, they go out of their way to opt back out—perhaps unsurprising given that for many people the question of organ donation ties into strongly held feelings about religion and spirituality and death in a way that participation in 401(k)s does not. (To be fair, there are significant methodological challenges to studying this issue. On the one hand, there are some problems with cross-national comparisons; on the other, as a pair of researchers pointed out in the *Journal of the Royal Society of Medicine* in 2014, if you restrict your analysis to a single country, comparing donation rates before and after a nudge is introduced, there's a chance that other variables could be responsible for any changes to the rate you observe.[37])

Some efforts to reduce consumers' energy use have delivered similarly unimpressive results. One nudge is to give consumers social feedback about how their electricity habits compare with those of their neighbors. It certainly makes sense to presume that if people realize they are an outlier in their neighborhood because they use too much energy, this could cause them to act in a more responsible manner. In practice, however, this doesn't always work. One paper from 2018 surveying past results notes that "large-scale randomized control trials in the United States have observed energy savings of around 2%, compared to a no-intervention control group . . . However, studies done in the United Kingdom, the Netherlands, and Australia found no statistically significant differences in energy consumption levels between households who had been given social comparison feedback and those who had not."[38] To be sure, a 2 percent reduction in energy might be more impressive than it sounds if it is generated by a low-cost tweak, but it's clear that some energy-use nudges haven't generated any return whatsoever.

Another major disappointment came in the form of "a five-year effort to design promising online and text-message interventions to improve college achievement through several distinct channels" that was published as a National Bureau of Economic Research working paper. In a large sample of almost twenty-five thousand students on three campuses, Philip Oreopoulos and Uros Petronijevic found "some improvement from coaching-based interventions on mental health and study time, but none of the interventions we evaluate significantly influences academic outcomes (even for those students more at risk of dropping out)."[39] A different NBER paper examined "the impact of a national and state-level campaign to encourage students to apply for financial aid" using nudge principles—one that reached a gigantic sample of more than eight hundred thousand students. That one found "no impacts on financial aid receipt or college enrollment overall or for any student subgroups," as well as "no evidence that different approaches to message framing, delivery, or timing, or access to one-on-one advising affected campaign efficacy."[40]

As for Obama's stimulus-cash nudge, it's a difficult thing to evaluate, but according to a working paper three economists published for the Federal Reserve Board, those who received the nudge version of their stimulus dollars were *less* likely to spend them than those who received them in a lump sum.[41] The researchers took advantage of the fact that in 2008 the Bush administration enacted a stimulus via lump-sum checks, and simply made the (admittedly apples-to-oranges) comparison between the two waves of stimulus, finding that "25 percent of households reported that the one-time economic stimulus payment in 2008 led them to mostly increase their spending while only 13 percent reported that the extra pay from the lower withholding in 2009 led them to mostly increase their spending." (This particular nudge also raises a fascinating issue about the politics of nudging: In shaping legislation to increase its effectiveness, Obama actually made it less likely the average American would understand they had received money from the government at all. Here is a perfect example of the "submerged state" concept laid out by the political scientist Suzanne Mettler, in which people don't comprehend the role the federal government plays in their life. A lump-sum check makes that role perfectly evident—as I am being reminded, as I write this, by President Trump's decision to slightly delay the issuance of coronavirus stimulus checks so as to ensure they are festooned with his signature.)

In short, there's nothing magical or infallible about nudging as a concept. Some nudges work, and others don't, and there is no stone tablet revealing, prior to pilot studies, which nudges will encounter which fate. Humans are complicated. We should all resist the pull of a certain flavor of too-blunt reasoning that often manifests itself in this style: *because humans are loss averse, an intervention that leverages loss aversion to change their behavior is likely to work*. Well, people are *generally* loss averse, but they're also many other things, which is why there's simply no guarantee that whatever lever is being utilized will be enough to change the behavior in question. (There's a difference between believing your kid likes coffee ice cream, as a general principle, and believing an offer of coffee ice cream will definitely get him to clean his room tonight.) One of the most common errors psychologists and laypeople alike make is to assume that just because some effect is uncovered in a lab, or in a specific real-world setting, it is broadly generalizable to other settings, too. Often, that isn't the case.

That said, the nudge revolution isn't just about the interventions themselves but also about certain changes to how organizations function: particularly, the institutionalization of experimentation. Thanks in large part to nudge-minded wonks who understand that slight tweaks to a program can yield sizable results, and that it's therefore important to test those different tweaks and see which ones are best, it is becoming increasingly common for traditionally sclerotic institutions to take a more experimental approach to what they do. New York City's attempts to improve its court-date nonresponse problem is one example, but there are numerous others: For example, the City of Chicago contracted with the University of Chicago, New York University, and ideas42 to help it understand how best to reduce plastic bag use. It tested different approaches, monitored the results, and came to the conclusion that a seven-cent tax "significantly reduced disposable bag use and increased reusable bag use."[42]

Of course, the idea of testing different approaches in this manner isn't new. Evaluating which of two messages has a bigger influence on people's behavior could be seen as an example of so-called A/B testing, which advertisers have been doing forever. But for other sorts of institutions, and particularly public-sector ones, this type of unflinching self-evaluation and willingness to swap out old approaches for new and better ones represents true progress.

VERY FEW PEOPLE WOULD disagree that all else being equal, it's better for institutions to operate with a sound understanding of human nature than with a naive one; that it's better to A/B test some new form than to skip that step; and that policies that take into account phenomena like loss aversion and the availability heuristic are better equipped to succeed than policies that ignore these facets of human nature. None of these elements of nudging is controversial.

What is controversial is the broader political context of nudging. Some critics of Sunstein in particular, perhaps most prominently Robert Kuttner, co-editor and cofounder of the liberal magazine *The American Prospect*, have pointed out that to focus too intently on nudging is to neglect bigger, more important issues. Nudge-style manipulation of people's decision-making "is highly creative as far as it goes," Kuttner wrote in one 2009 *Prospect* essay comparing nudging to traditional top-down "command and control" economic policies. "But to pursue the example of the employees helpfully 'nudged' into joining savings plans, the deeper problem today is that fewer companies offer pensions at all, and tax-deferred savings schemes such as 401(k)s (which aren't real pensions) are taking a beating from the stock-market collapse. Systemic reform requires more than a nudge; it may even require dreaded commands and controls like the expansion of Social Security."[43]

In a book review essay in *The Nation* of Sunstein's 2019 book *On Freedom*, the Yale historian and legal scholar Samuel Moyn issues a somewhat more philosophical critique; he notes the nudge godfather's failure to develop a "bigger theory of how people come by their desires, what forces stand in their way, and what democracy can do to help." Sunstein is very concerned with how people fail to correctly satisfy their preferences, or to properly weigh clashing preferences—I want to drink a soda; I want to stay healthy—in a sensible way. But should those preferences be taken as givens? What's missing from this account, Moyn argues, is that as a result of preexisting social pressure, advertising, and myriad other influences people can easily come to believe they want things that under different conditions they might not.

In other words, if you spend a lot of time and energy worrying about how the government can encourage people to eat a few more vegetables and drink fewer sodas without restricting people's freedom to drink sodas if they wish, it might distract you from questions such as why there is such a feverish market for artificially sweetened drinks in the first place, whether and to what extent people's seemingly limitless appetite for them is driven by a lack of affordable alternatives (or by government subsidies for corn, which is grown and processed into countless tons of corn syrup, which goes into soda), and so on. It's important to realize, in Moyn's account, that this desire does not materialize fully formed out of some sort of pure primordial pre-market mist, but is, like everything else, the product of decisions powerful people have made about how society should be organized. Tom Slee's 2006 book, *No One Makes You Shop at Wal-Mart: The Surprising Deceptions of Individual Choice*, while not about nudging per se, makes a similar sort of argument.[44] In it, Slee points out that ruthless market forces can lead to homogeneous, chain-store-dominated small towns that residents wouldn't actually choose if they had ever had true power to shape things. "Consumer choice" can't really displace a big-box store that has already demolished the competition. No one *makes* you shop at Walmart, but once America's political-economic system has gone to work on your town, you might have to drive an hour and a half to find any real alternative.

In Moyn's telling, Sunstein doesn't seem particularly curious about why society is structured the way it is; if people are beset with temptation and distraction, the best anyone in a position of power can really do is nudge them toward self-control. "Sunstein's obsession with self-control makes *On Freedom* often read like a book about piloting a boat through shark-infested waters that focuses mainly on the government's role in encouraging the passengers not to drink so much that they fall overboard," Moyn writes. "Sunstein is not interested in the sharks."[45] Moyn happens to be critiquing Sunstein from the left, but part of what makes his argument so compelling is that one could just as easily see a conservative making a similar case, albeit with different priorities in mind (say, with reference to people's preferences for pornography and endless online-dating options).

Moyn's essay brings to mind one of the strangest things about Sunstein's memoir, *Simpler*: its treatment of the subprime mortgage crisis. That crisis, alongside the fight over the Affordable Care Act, was the signature issue of Barack Obama's first term. And by the time *Simpler* was published, plenty of detailed, forensic accounts of what had gone wrong had been published. As William Simon noted in a 2013 *Boston Review* critique of Sunstein's book, "The most important causes of the recent crisis involved the externalization of risk by borrowers and lenders, ultimately to the federal government through its deposit insurance, lender-of-last-resort, and too-big-to-fail bailout practices."[46] These factors set off a chain reaction that devastated the economy, nearly causing a full-blown second Great Depression.

Sunstein was right there, in the White House, as the Obama administration sifted through the wreckage. Not only that, but he was serving as the most powerful regulator in the country. One would imagine he would have developed strong feelings about mortgages and banking practices and derivatives and the regulatory landscape surrounding them. And yet in *Simpler*, there is little discussion of the big, well-documented policy problems and deeply deceptive advertising that helped cause the crisis, or the seemingly obvious roles government could play in ameliorating these issues. Instead, it's almost all nudging when mortgages come up. At one point, for example, Sunstein approvingly cites "an effort to inform consumers by ensuring that mortgages are made simpler, shorter, and clearer."

Sure, that's all well and good: Mortgages should certainly be made consumer-friendly in their presentation. Elizabeth Warren, one of the United States' fiercest consumer advocates, has emphasized the point, and the government office she helped create, the Consumer Financial Protection Bureau, introduced a rule that created simpler, so-called Know Before You Owe mortgage paperwork in 2013.[47] But even if the fine print on mortgages prior to that reform made it easier for sellers to victimize buyers, "irrational consumer choices are not the major source of risk here," as Simon puts it.[48] The financial crisis was sparked by the fact that powerful institutions were allowed to sell financially dangerous products to vulnerable people. That is not a problem that can be nudged away; it isn't, at root, about individual consumers having made the "wrong" choices but about the choices that were available in the first place. Why shouldn't the government prevent consumers from "choosing" a mortgage they obviously can't afford, especially given that such mortgages are often guaranteed by the U.S. government, meaning taxpayers?

Sunstein may elevate nudging to a higher perch than it deserves, but some other thinkers in his orbit maintain a better sense of proportionality. A nice counterpoint to the myopia of *Simpler* comes in the fascinating 2020 book *Under the Influence: Putting Peer Pressure to Work*, by the economist Robert H. Frank. Frank is interested in behavioral-econ findings and even thanks Sunstein in his acknowledgments. But he also understands nudging's limitations, and the fact that it can't, at the end of the day, replace the necessary drudgery and challenges and political trade-offs of policy making of other sorts.

Take, for example, a section of his book about ameliorating "negative externalities"—the term economists use for any situation in which someone's actions cause harm to others and that harm isn't included in the cost of their performing that action. Pollution is the traditional textbook example: If my smokestack harms the respiratory health of nearby residents and I don't cover the cost of their increased health-care expenses, as well as the other public harms caused by the pollution my business is emitting (lowered economic productivity when people have to miss days of work, for example), I'm imposing a negative externality on society. When this is remedied via taxation, it's known as a Pigouvian tax, after Arthur Cecil Pigou, the British economist who invented the idea.

In Frank's view, nudging can get us only so far when it comes to solving a serious, complicated problem like the negative externalities polluters impose on society. At some point, you need to actually enact policy in the form of this type of tax:

> Given the astronomical rates of return being delivered by nudge units, governments have every reason to keep expanding them. But many of the biggest policy opportunities will continue to elude us unless we can discover ways of making Pigouvian tax remedies more palatable to voters. Invoking a social norm, for example, is effective in getting people to reuse their hotel towels, and putting healthful foods within easy reach is effective in getting more people to

> choose them. But when the stakes are considerably higher, more powerful incentives are often necessary.[49]

Now, one of the appeals of nudging is that it allows you to do *something* even when you have been cut off from more substantive options. An article in *The New Republic* described Obama's decision to pursue nudging via executive orders (such as setting up the nudge unit) in these terms: "For a presidency born in economic catastrophe and plagued by an anemic recovery, gross inequality, and a hostile Congress, there was always the question of how to use executive action to salvage something positive in the face of a hopeless political situation."[50]

But there's a difference between treating nudging as one limited tool among many others and embracing it as something of a full-blown ideology. Kuttner's and Moyn's pointed critiques of Sunstein, and Frank's gentler suggestion, all cut to the heart of the matter. They all helpfully expose the limits of nudging, and indeed of the broader Primeworld ethos. If the question is whether the best nudges can often meaningfully affect whether people choose option A versus option B in one particular instance, on one particular form, the answer is yes. If the question is whether nudging can, for example, meaningfully reverse the devastating effects of the great risk shift and solve the problems birthed by it, the answer is an emphatic no.

THIS LOGIC CERTAINLY APPLIES to the redesign of New York City's summons form. While that form became sleeker and easier to process as a result of the efforts of behavioral-science specialists, and while those who were summoned began getting nudged by text messages, the fact remained that the City of New York still assigned those accused of low-level offenses arbitrary court dates that posed severe logistical difficulties to low-income people, leading to all-too-predictable racial disparities, and still issued arrest warrants for those who failed to show up.

These problems were never going to be fully solved by nudging; rather, someone had to actually change the underlying policies themselves. This is as tidy an illustration of the limits of nudging as we are likely to find, and it was nicely expressed to me by Will Tucker, a young veteran of the behavioral science revolution who was a vice president at ideas42, and who helped cofound the White House's Social and Behavioral Sciences Team. "You can look at that study result and say, 'Whoa, we need to scale this result and send out a whole bunch more message reminders!' Or you can look at that and say, 'Wow—this data indicates the way our courts interact with our people is broken—we need to change how court dates work so that no one is thrown in Rikers because of a faded piece of carbon copy paper.'"

That's actually what happened. In 2016, New York's city council passed the Criminal Justice Reform Act. The legislation significantly shifted law-enforcement priorities and practices, most specifically by redefining many minor infractions as civil offenses. Rather than arresting a violator on the spot or issuing them a criminal summons form, the city responds to a civil offense by issuing a civil summons form. Importantly, a failure to appear in court to contest a civil fine doesn't lead to a bench warrant.

According to a City Council website touting the one-year anniversary of the reforms, there has been a massive reduction in the number of criminal summonses being issued. And, as that same site notes, "the Council also worked with four of the City's District Attorney offices to clear the summons backlog for minor offenses like having an open container or entering a park after hours. This dismissed over 644,000 outstanding warrants."[51]

As it turned out, I was myself a direct beneficiary of the Criminal Justice Reform Act. In August of 2019—I'm slightly ashamed to admit—I received a civil summons for public drinking (it was just myself and a friend in a riverside park, nothing remotely peace disturbing, I swear).[52] I subsequently forgot to pay the fine, and then, when I rediscovered the summons in my wallet months later—and months late—and called the number on it, I found out my case had been mysteriously tossed. Prior to the change in the law, I could have had a bench warrant issued for my arrest. And much more important, over the years many people in far more vulnerable situations than me *did* have such warrants issued, often leading to tremendous harm and disproportionate punishment for minor offenses and acts of forgetfulness.

As any advocate for criminal justice reform will tell you, New York City's criminal justice system still has a long way to go—the passage of the CJRA by no means ended the practices of unfairly or overzealously prosecuting vulnerable New Yorkers. But things *have* improved, and one of the key takeaways here is that while nudging is nice, and while low-cost tweaks within a given system can reap benefits, there's only so much you can do to improve things without changing the system itself. Updating the summons form made a significant difference, of course, but that difference was dwarfed by the passage of laws that yanked a great many minor cases of lawbreaking out of the "criminal" bucket altogether and dropped them into the "civil" one instead, leading to far more humane results.

You can't nudge your way out of a policy problem, in other words.

# CONCLUSION: ESCAPE FROM PRIMEWORLD

As we've seen, there are myriad reasons half-baked behavioral science catches on, and those reasons often have to do with the cultural or institutional context of a given idea—the problem it is attempting to solve, the societal currents it is riding, and so on. As we conclude this book, it's worth taking stock of the more general, less context-specific reasons why bad social science spreads and what the consequences might be, particularly when it comes to Primeworld accounts.

The simplest reasons half-baked ideas tend to prevail is that all else being equal, the human brain has an easier time latching onto simple and monocausal accounts than to complicated and multicausal ones.

Such accounts are more likely to be accepted as true and to spread. Our brains are built to be drawn to quick, elegant-seeming answers.

The legendary sociologist Charles Tilly nicely explains this in his account of human storytelling, *Why? What Happens When People Give Reasons . . . and Why*. He writes, "Stories provide simplified cause-effect accounts of puzzling, unexpected, dramatic, problematic, or exemplary events. Relying on widely available knowledge rather than technical expertise, they help make the world intelligible." Tilly calls storytelling "one of [the] great social interventions" of the human species, precisely because of its ability to simplify and boil down. But this is the same reason stories can lead us astray. "In our complex world, causes and effects always join in complicated ways," he writes. "Simultaneous causation, incremental effects, environmental effects, mistakes, unintended consequences, and feedback make physical, biological, and social processes the devil's own work—or the Lord's—to explain in detail. Stories exclude these inconvenient complications."[1]

Think of all the stories that have fueled half-baked psychology: "Soldiers can resist PTSD if their resilience is boosted"; "Women can close the workplace gender gap if they feel an enhanced sense of power"; "Poor kids can catch up to their richer peers if they develop more grit." In emphasizing one particular causal claim about deeply complicated systems and outcomes, these and the other blockbuster hits of contemporary psychology elide tremendous amounts of important detail.

It's likely that just as our brains prefer simple stories, within psychology, too, the professional incentives point toward the development of simpler rather than more complex theories. People who study human nature aren't immune to the siren call of simplicity. In a reply to one of her papers, the psychologist Nina Strohminger criticizes this tendency rather eloquently: "The fetishization of parsimony means that unwieldy theories are often dismissed on these grounds alone . . . No doubt there is something less satisfying about settling for inelegance, but the best theories won't always feel right. Elegance is not a suitable heuristic for veracity."[2] Scientists often have good reason to prefer parsimony—Occam's razor has its uses—but still: simple-seeming explanations of complex phenomena warrant skepticism.

Of course, simple and elegant and appealing theories are more likely to pay. If you're a psychologist in the twenty-first century, particularly a young one, you face a daunting landscape when it comes to making a name and therefore a career for yourself. Funding is being cut left and right, and the ongoing adjunctification of academia certainly hasn't spared psychology. There's one silver lining, though: the public is more interested in behavioral science than ever before. That's especially true if you can tell a simple, exciting, and above all *new* story about a subject of great societal concern.

This difficult environment undoubtedly influences how psychologists present their work to the public. In an insightful chapter in the 2017 book *The Politics of Social Psychology*, the social psychologists Hart Blanton and Elif Ikizer (you might remember Blanton as one of the more thoughtful and prolific critics of the implicit association test) posit the existence of "bullet-point bias." They define this bias simply as "the tendency [on the part of researchers] to advance diluted but provocative scientific conclusions in the media." As they explain, "By learning to communicate effectively to science reporters, through the editorial pages of national news sources and on the stages at TED, scientists might be able to charge higher speaking fees, pursue lucrative consulting jobs, secure book deals, and enjoy the perks of minor celebrity."[3]

Of course, if journalists were saying "no thanks" to what these scientists have on offer, there would be less of a problem. But "in a new media environment, where alternative news sources are ever a click away, media outlets have become increasingly reliant on material generated by scientists who know how to grab and retain audience interest. Science sells; more so the more easily it is communicated."[4]

These dynamics can also elevate self-justification over scientific justification, keeping dubious ideas in circulation. The social psychologist Carol Tavris, who with the late, great social psychologist Elliot Abramson coauthored an influential book about human folly and self-deception called *Mistakes Were Made (but Not by Me)*, summed this up to me nicely in an interview:

> Once you have committed yourself to a theory—and this is true of any of us—it becomes hard to accept criticisms of that theory, let alone evidence that you might be wrong about it. Scientists are not immune to this inclination, even though the whole nature of the scientific enterprise is to put your beliefs to the test—Is this what's going on here?—and see if the evidence supports it or not. But if you have also taken your theory into the public forum, you are now getting thousands upon thousands of dollars to educate people in companies and the government about your test, your measure, or your hugely popular idea, you now have a vested interest, financially, emotionally, and psychologically, in its being right. How easy is it going to be for you to say, "Maybe I went too far? Maybe I ignored the parts that didn't fit? Maybe this idea sounded appealing but has a few problems I didn't anticipate?" That, in terms of research and science, is the greatest danger of this TED-ification phenomenon: the impulse to oversimplify and cut around the edges.

The influence of a certain social class helps propel these ideas as well. This is captured elegantly in the journalist Anand Giridharadas's 2018 book, *Winners Take All: The Elite Charade of Changing the World*. Giridharadas's basic thesis is that a burgeoning, youngish, tech-savvy overclass has revolutionized philanthropy for the worse, focusing on forms of charity that don't lead to meaningful change. He describes "an ascendant power elite that is defined by the concurrent drives to do well and do good, to change the world while also profiting from the status quo. It consists of enlightened businesspeople and their collaborators in the worlds of charity, academia, media, government, and think tanks."[5]

Giridharadas argues that for reasons of self-interest and ideology (two concepts that often blur into each other), members of this elite are more interested in superficial, individually focused remedies to social problems than remedies that would involve the analysis of power and the redistribution of resources. In reality, there's a zero-sum nature to many human problems—if you give poor people more money, it has to be taken from someone else, perhaps by raising taxes on the rich—but these thought leaders avoid this fact by endlessly proposing win-win interventions that make everyone (well, everyone at a given high-powered conference or junket at least) feel good without upsetting the status quo.

Given the ever-growing power of what Daniel Drezner calls "the ideas industry"—all those TED Talks and intellectual-seeming conferences and junkets—Giridharadas's point is well-taken: all else being equal, the powerful people capable of throwing a lot of money at ideas, and turning people into stars by inviting them to high-profile speaking gigs, are likely to be very amenable to Primeworld accounts. If rich and powerful people like a certain type of idea, that's all the more reason to imagine that that type of idea will enjoy market advantages as a result.

MOST OF THE PEOPLE directly involved in Primeworld-style storytelling obtain benefits: researchers and their universities get attention from publications; publications get clicks from readers eager for the next life hack or surprising lab finding; researchers and universities enjoy the benefits of press coverage and seek more of it; media outlets understand the resonance of these types of stories and seek more of them; and so on. All the incentives in the cycle point toward overclaiming, oversimplifying, and brushing aside countervailing evidence.

But there are losers, too, of course, such as news consumers who develop false or oversimplified beliefs about the present state of psychological science and how it maps onto the real world. They may come to wrongly believe that, for example, grit is a powerful predictor of who does and doesn't succeed in life, or that if they could just feel a bit more powerful at work, they could overcome entrenched gender bias or other obstacles stymying their professional ascent.

Storytelling can also have unintended consequences in the long run, as the endless recitation of certain themes potentially shifts public understanding of how the world works. In their chapter, Blanton and Ikizer express concerns that credulous scientific coverage of so-called wise interventions[6]—interventions advertised as being capable of affecting behavior in fairly minimalist, low-cost ways—could change people's conception of inequality and oppression. If you bombard people with the idea that poor kids can do better in school simply by getting grittier, or that we can address racism simply by having more people take the IAT, opening their eyes to their implicit bias, then when these problems *don't* improve, it might cause consumers of pop science to blame vulnerable people for their own plights. As Blanton and Ikizer write, "Wise interventions, by drawing attention to 'simple' fixes, might make it easier for the public to imagine simple ways the disadvantaged could have undone their misfortunes, had they tried. This process might reduce sympathy and raise blame."[7]

Blanton and Ikizer ran a set of experiments to test this hypothesis using Amazon Mechanical Turk. The researchers exposed the Turk workers either to "actual media reports covering wise interventions designed to reduce racial inequalities (in education and health)" or to coverage of more comprehensive, complicated interventions geared toward similar ends. Overall, "those in the wise intervention condition more strongly endorsed the view that minority students can easily overcome relative disadvantages on their own, if they simply try." In addition, exposure to these interventions "not only promoted attributions of responsibility but also led to the assignment of blame" in the case of interventions targeting marginalized groups. These "increases in blame were strongest among participants who were politically conservative, as opposed to liberal."[8]

I would be missing the entire point of my own book if I argued that a small group of tightly controlled online experiments *proves* that exposure to Primeworld innovations will cause people to feel less empathy toward the disadvantaged or be more likely to blame them for their plight. But it does provide some evidence pointing in that direction. It's worth asking, given the popularity of Primeworld and the shallowness of many of its scientific claims, just what all this might do to our collective notions of justice and fairness and social reform in the long run.

Blanton and Ikizer aren't the only psychologists to have probed this question empirically. In one article in the *Journal of Experimental Social Psychology*, Natalie Daumeyer, then a Yale University doctoral student, and her colleagues found that in four experiments "perpetrators of discrimination are held less accountable and often seen as less worthy of punishment when their behavior is attributed to implicit rather than to explicit bias. Moreover, at least under some circumstances, people express less concern about, and are less likely to support efforts to combat, implicit compared with explicit bias."[9]

So by promoting the oversimplified assumption that a great deal of the world's discriminatory outcomes are the result of implicit bias, psychologists and other storytellers could be nudging people's broader beliefs on that subject in one particular direction, and not necessarily a desirable one.

THE CORONAVIRUS PANDEMIC of 2020 hit while I was finishing up this book. It's always a bad idea, when faced with an emotionally charged crisis, to become convinced that that crisis proves you were right all along. Catastrophes breed confirmation bias. And yet I have to admit that as I've contemplated the crisis, I've grown only more jaded about Primeworld; I have become only more convinced that it is a fundamentally impoverished account of society.

When the virus arrived in the United States, Americans' choices were, as always, defined by big, complicated structures of power and wealth. Some Americans in the pandemic's epicenters were forced to choose between financial ruin and continuing to work low-wage jobs in which they faced infection, while others were able to make a fairly seamless shift to working remotely. It would be impossible to overstate the significance of these divergences: no, having money didn't render anyone immune from the virus, but overall one's chance of riding this pandemic out safely and in relative comfort had everything to do with the resources at one's disposal, which, as usual, meant that shocking racial disparities soon emerged. Structural forces went a long way toward dictating who lived, who died, who struggled, and who was relatively unaffected. They always do.

The differences in individual behavior that arose during this period, too, seemed to have almost nothing to do with Primeworld and a great deal to do with the potent influence of social and political and religious networks. Some cities and states closed down quickly, rendering moot many individual choices on matters of social distancing; others dragged their feet interminably. While New York City's nonessential businesses were shut down, news outlets were broadcasting live video of spring break partiers packing beaches in Florida (and New York City was seen to have responded slower than some other cities, like San Francisco). Even within geographic areas, there were staggering differences in how people processed the news and subsequently changed their behavior, or didn't. In Park Slope, Brooklyn, which is full of well-off, highly educated families, those who owned or could rent summer homes in the country fled the city as quickly as possible; a few subway stops away, in the Hasidic parts of Crown Heights and Williamsburg, community halls were still packed, shoulder to shoulder, for weddings.

Whether and how quickly an individual reacted to the coronavirus news was clearly determined by their ties to other human beings, and the choices available to them were determined by the resources they had at their disposal. This shouldn't be a surprise. Of course, as the nudge chapter showed, there are ways in which tweaks to public-health messaging and various renovations to choice architecture can affect people's behavior at the margins. There's a place for these subtle interventions. But during a nationwide disaster, the sheer marginality of their usefulness becomes only more apparent.

Some psychologists see the present limitations of their field clearly. In April 2020, a group of them did something that would have been unthinkable not long ago: they publicly argued that people should *not* be listening to psychologists at that tumultuous moment, because the field was simply not ready for prime time.

This statement was made in a paper titled "Is Social and Behavioural Science Evidence Ready for Application and Dissemination?," published online as a preprint, meaning it was posted online before it went through the full peer-review and editing process. In it, a team led by Hans IJzerman, Andrew Przybylski, and Neil Lewis Jr., which also included other highly respected social psychologists like Simine Vazire (herself an outspoken open-science advocate), Patrick Forscher, and Stuart Ritchie, pointed out the temptation for behavioral scientists to apply their findings, in confident ways, to major issues of real-world importance like the pandemic.

But it was a mistake for psychology to be offering these services at that moment, they argued. While the field had made progress and was in the process of reforming itself, there was reason to be skeptical that "psychology is mature enough to provide" useful insights on "life-and-death issues like a pandemic." After making some suggestions about how to more effectively communicate uncertainty in psychological evidence, they struck a humble final note: "We believe that, rather than appealing to policy-makers to recognise our value, we should focus on earning the credibility that legitimates a seat at the policy table."[10]

Think about how many incentives point against publishing a paper like "Is Social and Behavioural Science Evidence Ready for Application and Dissemination?" Think how many professional opportunities a psychologist is giving up by acknowledging that his or her field is in no condition to sell its wares to a public with an insatiable appetite for behavioral science answers.

Of course, this paper doesn't necessarily represent the majority opinion of psychologists. As we've seen, there is still some reluctance to acknowledge the full scale of the field's problems. So it would be wrong to depict psychology's progress toward reform with a straight line, to pretend away the ongoing existence of myriad incentives nudging scientists to overclaim, university press offices to overhype, and exhausted journalists to accept and communicate clickbait-level "findings" at face value. These problems remain real and in some cases could be exacerbated as the pandemic delivers harsh financial blows to already-struggling research institutions and journalism outlets alike.

But a paper like the one IJzerman and his colleagues published is a sure sign that things are changing—that some psychologists, at least, are exhibiting a level of humility and realism about the complexity of their work that they have all too often lacked in recent years. It's a sign that in twenty or thirty years, if all goes well, a book like this one will no longer be necessary.

# NOTES

INTRODUCTION

1. Gregory Mitchell, "Jumping to Conclusions: Advocacy and Application of Psychological Research," Virginia Public Law and Legal Theory Research Paper, no. 2017–31 (May 2017): 139, ssrn.com/abstract=2973892.
2. Jesse Singal, "Psychology's Favorite Tool for Measuring Racism Isn't Up to the Job," *The Cut*, January 11, 2017, thecut.com/2017/01/psychologys-racism-measuring-tool-isnt-up-to-the-job.html. (Originally published on the Science of Us vertical, which was later folded into The Cut.)
3. Daniel T. Rodgers, *Age of Fracture* (Cambridge, Mass.: Belknap Press, 2011), 3.
4. For transparency's sake: There are some similarities between my Primeworld concept and the idea of MarketWorld introduced by Anand Giridharadas in his great 2018 book, *Winners Take All: The Elite Charade of Changing the World*, which I reference in this book's concluding chapter. I believe we developed the concepts independently. The earliest record

of my own usage of "Primeworld" is a May 13, 2018, email to a psychology researcher in which I raise the idea of perhaps calling this concept I am referencing "PrimeWorld," and Giridharadas's book didn't come out until more than three months after that. When I emailed Giridharadas in 2019 to ask when he had first used the term publicly, he replied, "I think the first public airing of MarketWorld would have been just before the book came out last summer, but I may have used it in talks or something before that—I can't be sure." All of which suggests we did develop the concepts independently, without eliminating the possibility that I somehow caught wind of his idea and it influenced the development of mine.

## 1. THE SELLING OF SELF-ESTEEM

1. David E. Early, "John B. Vasconcellos: 38 Years a Human Potential Titan in Sacramento, Dead at 82," *Mercury News*, May 24, 2014, mercurynews.com/2014/05/24/john-b-vasconcellos-38-years-a-human-potential-titan-in-sacramento-dead-at-82.
2. Will Storr, *Selfie: How We Became So Self-Obsessed and What It's Doing to Us* (New York: Harry N. Abrams, 2018), 190.
3. Storr, *Selfie*, 175.
4. Nathaniel Branden, *The Psychology of Self-Esteem: A Revolutionary Approach to Self-Understanding That Launched a New Era in Modern Psychology* (San Francisco: Jossey-Bass, 2001), 109.
5. Nathaniel Branden, "Life & Liberty: In Defense of Self," *Reason*, November 1, 1984.
6. Morris Rosenberg, *Society and the Adolescent Self-Image* (Princeton, N.J.: Princeton University Press, 1965).
7. Storr, *Selfie*, 191.
8. Garry Trudeau, *Doonesbury*, February 19, 1990, available at: www.gocomics.com/doonesbury/1990/02/19.
9. Jay Mathews, "Picking Up a Head of Esteem," *The Washington Post*, February 20, 1988, washingtonpost.com/archive/lifestyle/1988/02/20/picking-up-a-head-of-esteem/4f55a1aa-21a3–4a9e-9b7f-5e7444ad4254.
10. Associated Press, "Lampooned California Commission Gets Some Respect," December 31, 1988.
11. Benjamin Franklin, *Poor Richard's Almanack* (New York: Barnes & Noble, 2004), 536, Kindle edition.
12. Alfred Whitney Griswold, "New Thought: A Cult of Success," *American Journal of Sociology* 40, no. 3 (November 1934): 309–18, jstor.org/stable/2768263.
13. Frank Channing Haddock, "The Second Lesson—the Mood of Success," in *The King's Achievements; or, Power for Success Through Culture of Vibrant Magnetism* (Lynn, Mass.: Nichols Press, 1903), 10, books.google.com/books?id=ddNVW82xz-0C.
14. Norman Vincent Peale, *The Power of Positive Thinking* (Noida, India: Om Books International, 2016), 152.

15. P. Shah, "Dreams Do Come True, Got My Dream Car!," *The Secret*, thesecret.tv/stories/dreams-come-true-got-dream-car.
16. "Law of Attraction," *The Secret*, thesecret.tv/law-of-attraction.
17. William James, *The Principles of Psychology*, vol. 1 (New York: Holt, 1890), psychclassics.yorku.ca/James/Principles/prin10.htm.
18. Storr, *Selfie*, 199.
19. Storr, *Selfie*, 200.
20. Storr, *Selfie*, 201.
21. Storr, *Selfie*, 211.
22. Storr, *Selfie*, 210.
23. Storr, *Selfie*, 206.
24. Storr, *Selfie*, 214.
25. Storr, *Selfie*, 215.
26. Storr, *Selfie*, 214–5; Jesse Singal, "How the Self-Esteem Craze Took Over America," *The Cut*, May 30, 2017, thecut.com/2017/05/self-esteem-grit-do-they-really-help.html.
27. California State Department of Education, *Toward a State of Esteem: The Final Report of the California Task Force to Promote Self-Esteem and Personal and Social Responsibility* (Sacramento: California State Department of Education, 1990), files.eric.ed.gov/fulltext/ED321170.pdf.
28. Storr, *Selfie*, 202.
29. United Press International, "Self-Esteem Task Force Comes on Strong," March 25, 1987.
30. Ellen Uzelac, "The I'm-OK, You're-OK Task Force," *Baltimore Sun*, March 20, 1989.
31. "The Market for Self-Improvement Products and Services," PR Newswire, January 20, 2015, prnewswire.com/news-releases/the-market-for-self-improvement-products—services-289121641.html.
32. Lena Williams, "Using Self-Esteem to Fix Society's Ills," *The New York Times*, March 28, 1990.
33. Vivian Smith, "California Magic," *Globe and Mail*, June 15, 1990.
34. John Rosemond, "Kids Aren't Special; Their Actions Are," *Buffalo News*, October 4, 1999, buffalonews.com/1999/10/04/kids-arent-special-their-actions-are.
35. Will Storr, "The Man Who Destroyed America's Ego," *Matter* (blog), *Medium*, February 25, 2014, medium.com/matter/the-man-who-destroyed-americas-ego-94d214257b5.
36. Roy F. Baumeister, Dianne M. Tice, and Debra G. Hutton, "Self-Presentational Motivations and Personality Differences in Self-Esteem," *Journal of Personality* 57, no. 3 (September 1989): 547–79, doi.org/10.1111/j.1467–6494.1989.tb02384.x.
37. Roy F. Baumeister et al., "Exploding the Self-Esteem Myth," *Scientific American*, December 2005, scientificamerican.com/article/exploding-the-self-esteem-2005–12.
38. Roy F. Baumeister et al., "Does High Self-Esteem Cause Better Performance, Interpersonal Success, Happiness, or Healthier Lifestyles?," *Psychological Science in the Public Interest* 4, no. 1 (2003): 1–44, doi.org/10.1111/1529–1006.01431.

39. I used this as a source for the average height for basketball players: probasketballtroops.com/average-nba-height. And this for the average height for American males: halls.md/average-height-men-height-weight.
40. Baumeister et al., "Does High Self-Esteem Cause Better Performance?," 12; E. M. Skaalvik and K. A. Hagtvet, "Academic Achievement and Self-Concept: An Analysis of Causal Predominance in a Developmental Perspective," *Journal of Personality and Social Psychology* 58, no. 2 (February 1990): 292–307, doi.org/10.1037/0022–3514.58.2.292.
41. Baumeister et al., "Does High Self-Esteem Cause Better Performance?," 11. See also J. G. Bachman and P. M. O'Malley, "Self-Esteem in Young Men: A Longitudinal Analysis of the Impact of Educational and Occupational Attainment," *Journal of Personality and Social Psychology* 35, no. 6 (1977): 365–80, doi.org/10.1037//0022–3514.35.6.365.
42. Baumeister et al., "Does High Self-Esteem Cause Better Performance?," 11.
43. Baumeister et al., "Does High Self-Esteem Cause Better Performance?," 11.
44. Baumeister et al., "Does High Self-Esteem Cause Better Performance?," 13.
45. Baumeister et al., "Does High Self-Esteem Cause Better Performance?," 4.
46. Her archives make it clear she was not a fan. See drlaurablog.com/tag/self-esteem.
47. Carol S. Dweck, *Mindset: The New Psychology of Success* (New York: Ballantine Books, 2016), ix.
48. David S. Yeager et al., "A National Experiment Reveals Where a Growth Mindset Improves Achievement," *Nature* 573 (September 2019): 364, doi.org/10.1038/s41586–019–1466-y.
49. Yeager, "A National Experiment Reveals Where a Growth Mindset Improves Achievement."
50. Stefan G. Hofmann et al., "The Efficacy of Cognitive Behavioral Therapy: A Review of Meta-analyses," *Cognitive Therapy and Research* 36, no. 5 (October 2012): 427–440, doi.org/10.1007/s10608–012–9476–1.
51. Storr, *Selfie*, 204.
52. Mass. Gen. Laws ch. 69, § 1.
53. Robert Salladay, "In Sacramento, a Revival of the Self-Esteem Movement," *San Francisco Chronicle*, September 7, 2001, sfgate.com/news/article/In-Sacramento-a-revival-of-the-self-esteem-2880073.php.
54. Chronicle Staff Report, "County Guaranteed Voice on SFO Expansion Plans / Davis OKs Bill Pushed by Peninsula Officials," *San Francisco Chronicle*, October 6, 2001, sfgate.com/politics/article/County-guaranteed-voice-on-SFO-expansion-plans-2872719.php.
55. D. Tambo and David C. Gartrell, "Preliminary Guide to the John Vasconcellos Papers," Department of Special Collections, Davidson Library, University of California, Santa Barbara, June 15, 2004, oac.cdlib.org/findaid/ark:/13030/kt3c601926/entire_text.
56. Eric Bailey, "Giving New Meaning to 'Youth Vote,'" *Los Angeles Times*, March 9, 2004, latimes.com/archives/la-xpm-2004-mar-09-me-voteage9-story.html; Rio Bauce, "Transferring Teen Clout to the Voting Booth," *San Francisco Chronicle*, April 17, 2015, sfgate.com/opinion/article/YOUTH-PERSPECTIVE-Transferring-teen-clout-to-2679610.php.

## 2. THE SUPERPREDATORS AMONG US

1. Ashley Nellis, *A Return to Justice: Rethinking Our Approach to Juveniles in the System* (Lanham, Md.: Rowman & Littlefield, 2015), 33–36.
2. James Alan Fox, *Trends in Juvenile Violence: A Report to the United States Attorney General on Current and Future Rates of Juvenile Offending*, Bureau of Justice Statistics (1996), bjs.gov/content/pub/pdf/tjvfox2.pdf.
3. James Alan Fox and Marianne W. Zawitz, "Homicide Trends in the United States," Bureau of Justice Statistics, bjs.gov/content/pub/pdf/htius.pdf.
4. Jon Jeter, "Youth Gets 2 Life Terms in Baby Killing," *The Washington Post*, October 27, 1993, washingtonpost.com/archive/local/1993/10/27/youth-gets-2-life-terms-in-baby-killing/1d4c8306-f889–4387-bbe9–28aa7d28fa41.
5. Evelyn Nieves, "3 Teen-Agers Are Sentenced in Strangling," *The New York Times*, June 9, 1992, nytimes.com/1992/06/09/nyregion/3-teen-agers-are-sentenced-in-strangling.html.
6. Associated Press, "Boy, 5, Is Killed for Refusing to Steal Candy," *The New York Times*, October 15, 1994, nytimes.com/1994/10/15/us/boy-5-is-killed-for-refusing-to-steal-candy.html.
7. "Yummy Sandifer's Revenge," *Chicago Tribune*, August 31, 2004, chicagotribune.com/news/ct-xpm-2004–08–31–0408310094-story.html.
8. Nancy R. Gibbs, "Murder in Miniature," *Time*, June 24, 2001, content.time.com/time/magazine/article/0,9171,165100,00.html.
9. Don Terry, "Boy Sought in Teen-Ager's Death Is Latest Victim of Chicago Guns," *The New York Times*, September 2, 1994, nytimes.com/1994/09/02/us/boy-sought-in-teen-ager-s-death-is-latest-victim-of-chicago-guns.html.
10. Craig Perkins and Patsy Klaus, *Criminal Victimization 1994*, U.S. Department of Justice Bureau of Justice Statistics (Washington, D.C.: U.S. Department of Justice, 1996), bjs.gov/content/pub/pdf/Cv94.pdf.
11. David E. Pitt, "Jogger's Attackers Terrorized at Least 9 in 2 Hours," *The New York Times*, April 22, 1989, nytimes.com/1989/04/22/nyregion/jogger-s-attackers-terrorized-at-least-9-in-2-hours.html.
12. Leonard Greene, "Trump Called for Death Penalty After Central Park Jogger Attack, and Still Has No Sympathy for Accused Despite Convictions Overturned," New York *Daily News*, July 19, 2018, nydailynews.com/new-york/ny-news-trump-death-penalty-central-park-five-20180713-story.html.
13. Pitt, "Jogger's Attackers Terrorized at Least 9 in 2 Hours."
14. Michael Welch, Eric A. Price, and Nana Yankey, "Moral Panic over Youth Violence: Wilding and the Manufacture of Menace in the Media," *Youth and Society* 34, no. 1 (September 2002): 3–30, doi.org/10.1177/0044118X02034001001.
15. J. Anthony Lukas, "Wilding—as American as Tom Sawyer," *The New York Times*, May 28, 1989, nytimes.com/1989/05/28/opinion/wilding-as-american-as-tom-sawyer.html.
16. James Traub, "The Criminals of Tomorrow," *The New Yorker*, November 4, 1996, 50. (emphasis in original)

17. Mary Ann Meyers, "John DiIulio Gets Religion," *Pennsylvania Gazette*, September 29, 1997, upenn.edu/gazette/1097/dool.html.
18. John J. DiIulio Jr., *Governing Prisons: A Comparative Study of Correctional Management* (New York: Free Press, 1987).
19. National Center for Juvenile Justice, *Juvenile Offenders and Victims: 2014 National Report*, ed. Melissa Sickmund and Charles Puzzanchera (Pittsburgh: National Center for Juvenile Justice, 2014), ojjdp.gov/ojstatbb/nr2014/downloads/chapter1.pdf.
20. Jeffery T. Ulmer and Darrell Steffensmeier, "The Age and Crime Relationship: Social Variation, Social Explanations," in *The Nurture Versus Biosocial Debate in Criminology: On the Origins of Criminal Behavior and Criminality*, ed. Kevin M. Beaver, James C. Barnes, and Brian B. Boutwell (London: SAGE, 2014), 377–96, dx.doi.org/10.4135/9781483349114.n24.
21. Central Intelligence Agency, Directorate of Intelligence, *The Youth Bulge: A Link Between Demography and Instability* (Washington, D.C., 1986), cia.gov/library/readingroom/docs/CIA-RDP97R00694R000500680001–1.pdf.
22. Eli Lehrer, "The Real John DiIulio," Heritage Foundation, February 7, 2001, heritage.org/crime-and-justice/commentary/ed020701b-the-real-john-dilulio.
23. John DiIulio, "The Coming of the Super-Predators," *Weekly Standard*, November 27, 1995, washingtonexaminer.com/weekly-standard/the-coming-of-the-super-predators.
24. Carole Marks, "The Urban Underclass," *Annual Review of Sociology* 17, no. 1 (August 1991): 445–66, doi.org/10.1146/annurev.so.17.080191.002305.
25. John J. DiIulio Jr., "My Black Crime Problem, and Ours," *City Journal* (Spring 1996): 14–28, city-journal.org/html/my-black-crime-problem-and-ours-11773.html.
26. William J. Bennett, John J. DiIulio, and John P. Walters, *Body Count: Moral Poverty . . . and How to Win America's War Against Crime and Drugs* (New York: Simon & Schuster, 1996), 21.
27. Jacques Steinberg, "The Coming Crime Wave Is Washed Up," *The New York Times*, January 3, 1999, nytimes.com/1999/01/03/weekinreview/ideas-trends-storm-warning-the-coming-crime-wave-is-washed-up.html.
28. *Newsweek* Staff, "'Superpredators' Arrive," *Newsweek*, January 21, 1996, newsweek.com/superpredators-arrive-176848.
29. Richard Zoglin, "Now for the Bad News: A Teenage Time Bomb," *Time*, January 15, 1996, content.time.com/time/magazine/article/0,9171,983959,00.html.
30. DiIulio, "The Coming of the Super-Predators"; Richard B. Freeman, "Who Escapes? The Relation of Churchgoing and Other Background Factors to the Socioeconomic Performance of Black Male Youths from Inner-City Tracts," in *The Black Youth Employment Crisis*, ed. Richard B. Freeman and Harry J. Holzer (Chicago: University of Chicago Press, 1986), 373.
31. John J. DiIulio, Jr., "Supporting Black Churches: Faith, Outreach, and the Inner-City Poor," *Brookings Review* (Spring 1999), 44.

32. "Youth Perpetrators of Serious Violent Crime," ChildStats.gov, www.childstats.gov/americaschildren/beh5.asp.
33. See, for example, Jeff Grogger and Michael Willis, "The Emergence of Crack Cocaine and the Rise in Urban Crime Rates," *The Review of Economics and Statistics* 82, no. 4 (November 2000): 519–29, doi.org/10.1162/003465300558957; Alfred Blumstein, "Youth Violence, Guns, and the Illicit-Drug Industry," *Journal of Criminal Law and Criminology* 86, no. 1 (Fall 1995): 10–36.
34. James Q. Wilson, *Crime*, ed. Joan Petersilia (San Francisco: ICS Press, 1995).
35. James C. Howell, *Preventing and Reducing Juvenile Delinquency: A Comprehensive Framework* (Thousand Oaks, Calif.: SAGE, 2009), 6, dx.doi.org/10.4135/9781452274980.
36. E. G. M. Weitekamp et al., "On the 'Dangerousness' of Chronic/Habitual Offenders: A Re-Analysis of the 1945 Philadelphia Birth Cohort," *Studies on Crime and Crime Prevention* 4, no. 2 (1995): 159–75, ncjrs.gov/App/publications/abstract.aspx?ID=158807.
37. Thomas Nagel, *Mortal Questions* (Cambridge, UK: Cambridge University Press, 1979), 34.
38. Daniel W. Drezner, *The Ideas Industry: How Pessimists, Partisans, and Plutocrats Are Transforming the Marketplace of Ideas* (New York: Oxford University Press, 2017).
39. Nellis, *Return to Justice*, 42–43.
40. *Hearings on the Juvenile Justice and Delinquency Act, Before the Subcommittee on Early Childhood, Youth, and Families*, 104th Cong. (1996).
41. Richard Sisk, "N.Y. Faces Teen Crime 'Time Bomb'—Study," New York *Daily News*, February 25, 1996, nydailynews.com/archives/news/n-y-faces-teen-crime-time-bomb-study-article-1.719954.
42. Hillary Rodham Clinton, "1996: Hillary Clinton on 'Superpredators,'" C-SPAN, February 25, 2016, YouTube, youtu.be/j0uCrA7ePno.
43. Andrew Kaczynski, "Biden in 1993 Speech Pushing Crime Bill Warned of 'Predators on Our Streets' Who Were 'Beyond the Pale,'" *CNN*, March 7, 2019, cnn.com/2019/03/07/politics/biden-1993-speech-predators/index.html.
44. Mark Kleiman, "The Current Crime Debate Isn't Doing Hillary Justice," *Washington Monthly*, February 17, 2016, washingtonmonthly.com/2016/02/17/the-current-crime-debate-isnt-doing-hillary-justice. Thanks to Nathan J. Robinson for pulling together these examples in his book *Superpredator: Bill Clinton's Use and Abuse of Black America* (West Somerville, Mass.: Current Affairs Press, 2016).
45. Michael Bochenek, "Trying Children in Adult Courts," in *No Minor Matter: Children in Maryland's Jails* (New York: Human Rights Watch, 1999).
46. Nellis, *Return to Justice*, 39.
47. Nellis, *Return to Justice*, 39.
48. Franklin E. Zimring, "American Youth Violence: A Cautionary Tale," ed. Michael Tonry, *Crime and Justice in America 1975–2015*, 42, no. 1 (2013): 265–98.

49. Mary Gibson, *Born to Crime: Cesare Lombroso and the Origins of Biological Criminology* (Westport, Conn.: Praeger, 2002), 19–20.
50. Gibson, *Born to Crime*, 100.
51. Gibson, *Born to Crime*, 101.
52. Michael T. Light and Jeffery T. Ulmer, "Explaining the Gaps in White, Black, and Hispanic Violence Since 1990: Accounting for Immigration, Incarceration, and Inequality," *American Sociological Review* 81, no. 2 (April 2016): 290, doi:10.1177/0003122416635667.
53. Dionissi Aliprantis, "Human Capital in the Inner City," *Empirical Economics* 53, no. 3 (November 2017): 1125–69, doi.org/10.1007/s00181–016–1160-y.
54. Khalil Gibran Muhammad, *The Condemnation of Blackness: Race, Crime, and the Making of Modern Urban America* (Cambridge, Mass.: Harvard University Press, 2010), 23.
55. N. S. Shaler, "Science and the African Problem," *The Atlantic*, July 1890, theatlantic.com/magazine/archive/1890/07/science-and-the-african-problem/523647.
56. Muhammad, *The Condemnation of Blackness*, 18.
57. Frederick L. Hoffman, *Race Traits and Tendencies of the American Negro* (New York: Macmillan, 1896), 17.
58. See, for example, Geoff K. Ward, *The Black Child-Savers: Racial Democracy and Juvenile Justice* (Chicago: University of Chicago Press, 2012).
59. Peter Wallsten, Tom Hamburger, and Nicholas Riccardi, "Bush Rewarded by Black Pastors' Faith," *Los Angeles Times*, January 18, 2005, latimes.com/archives/la-xpm-2005-jan-18-na-faith18-story.html.
60. Gustav Niebuhr, "A Point Man for the Bush Church-State Collaboration," *The New York Times*, April 7, 2001, nytimes.com/2001/04/07/us/public-lives-a-point-man-for-the-bush-church-state-collaboration.html.
61. Remedies Brief of Amici Curiae Jeffrey Fagan et al., *Kuntrell Jackson v. State of Arkansas*, Civil Action no. 10–9647 (DDC filed January 17, 2012), eji.org/files/miller-amicus-jeffrey-fagan.pdf.
62. Adam Liptak and Ethan Bronner, "Justices Bar Mandatory Life Terms for Juveniles," *The New York Times*, June 25, 2012, nytimes.com/2012/06/26/us/justices-bar-mandatory-life-sentences-for-juveniles.html; "United States Supreme Court Juvenile Justice Jurisprudence," National Juvenile Defender Center, njdc.info/practice-policy-resources/united-states-supreme-court-juvenile-justice-jurisprudence.
63. Elizabeth Becker, "As Ex-theorist on Young 'Superpredators,' Bush Aide Has Regrets," *The New York Times*, February 9, 2001, nytimes.com/2001/02/09/us/as-ex-theorist-on-young-superpredators-bush-aide-has-regrets.html.

## 3. OF POSING AND POWER

1. Spencer Johnson, *Who Moved My Cheese? An Amazing Way to Deal with Change in Your Work and in Your Life* (New York: G. P. Putnam's Sons, 1998).
2. Dana R. Carney, Amy J. C. Cuddy, and Andy J. Yap, "Power Posing: Brief Nonverbal Displays Affect Neuroendocrine Levels and Risk

Tolerance," *Psychological Science* 21, no. 10 (2010): 1363–68, doi.org/10.1177/0956797610383437.

3. Amy Cuddy, "Amy Cuddy: Passing as Myself," Facebook, February 11, 2017, facebook.com/amycuddycommunity/posts/705005856338544.
4. Amy Cuddy, "The Story Collider," SoundCloud, 2014, soundcloud.com/the-story-collider/amy-cuddy-passing-as-myself.
5. David Hochman, "Amy Cuddy Takes a Stand," *The New York Times*, September 19, 2014, nytimes.com/2014/09/21/fashion/amy-cuddy-takes-a-stand-TED-talk.html.
6. "Amy Cuddy Strikes a Power Pose," CBS News, December 13, 2015, cbsnews.com/news/amy-cuddy-strikes-a-power-pose.
7. Chen-Bo Zhong and Katie Liljenquist, "Washing Away Your Sins: Threatened Morality and Physical Cleansing," *Science* 313, no. 5792 (September 2006): 1451–52, doi.org/10.1126/science.1130726.
8. Brian D. Earp et al., "Out, Damned Spot: Can the 'Macbeth Effect' Be Replicated?," *Basic and Applied Social Psychology* 36, no. 1 (2014): 91–98, doi.org/10.1080/01973533.2013.856792.
9. Daniel Engber, "Sad Face," *Slate*, August 28, 2016, slate.com/articles/health_and_science/cover_story/2016/08/can_smiling_make_you_happier_maybe_maybe_not_we_have_no_idea.html.
10. Keith M. Welker, "Upright and Left Out: Posture Moderates the Effects of Social Exclusion on Mood and Threats to Basic Needs," *European Journal of Social Psychology* 43, no. 5 (May 2013): 355–61, doi.org/10.1002/ejsp.1944.
11. Vanessa K. Bohns and Scott Wiltermuth, "It Hurts When I Do This (or You Do That): Posture and Pain Tolerance," *Journal of Experimental Social Psychology* 48, no. 1 (2012): 341–45, digitalcommons.ilr.cornell.edu/articles/1077.
12. Eva Ranehill et al., "Assessing the Robustness of Power Posing," *Psychological Science* 26, no. 5 (2015): 653–56, doi.org/10.1177/0956797614553946.
13. Andrew Gelman and Kaiser Fung, "The Power of the 'Power Pose,'" *Slate*, January 19, 2016, slate.com/articles/health_and_science/science/2016/01/amy_cuddy_s_power_pose_research_is_the_latest_example_of_scientific_overreach.html.
14. Dana Carney, "My Position on 'Power Poses,'" University of California, Berkeley, faculty.haas.berkeley.edu/dana_carney/pdf_my%20position%20on%20power%20poses.pdf.
15. Andrew Gelman and Deborah Nolan, "You Can Load a Die, But You Can't Bias a Coin," *The American Statistician* 56, no. 4 (2002): pp. 308–311, https://doi.org/10.1198/000313002605.
16. Joseph P. Simmons and Uri Simonsohn, "Power Posing: P-Curving the Evidence," *Psychological Science* 28, no. 5 (2017): 687–93, doi.org/10.1177/0956797616658563.
17. Quentin F. Gronau et al., "A Bayesian Model–Averaged Meta-analysis of the Power Pose Effect with Informed and Default Priors: The Case of Felt Power," *Comprehensive Results in Social Psychology* 2, no. 1 (June 2017): 123–38, doi.org/10.1080/23743603.2017.1326760.

18. Amy J. C. Cuddy, S. Jack Schultz, and Nathan E. Fosse, "*P*-Curving a More Comprehensive Body of Research on Postural Feedback Reveals Clear Evidential Value for Power-Posing Effects: Reply to Simmons and Simonsohn (2017)," *Psychological Science* 29, no. 4 (April 2018): 656, doi.org/10.1177/0956797617746749. (emphasis in original)
19. Joe Simmons, Leif Nelson, and Uri Simonsohn, "Outliers: Evaluating a New P-Curve of Power Poses," *Data Colada* (blog), December 6, 2017, datacolada.org/66.
20. Marcus Crede, "A Negative Effect of a Contractive Pose Is Not Evidence for the Positive Effect of an Expansive Pose: Comment on Cuddy, Schultz, and Fosse (2018)," *Meta-psychology* 3 (2019), doi.org/10.15626/mp.2019.1723.
21. "Jaylen Brown Full Game 6 Highlights vs Cavaliers 2018 Playoffs ECF-27 Points!," House of Highlights, May 25, 2018, YouTube, youtube.com/watch?v=-qylB2qUKKU.
22. Erving Goffman, *The Presentation of Self in Everyday Life* (Edinburgh, UK: University of Edinburgh, 1956), 34.
23. Jesse Singal, "The 5-Step Approach to Not Being Misunderstood Anymore," *The Cut*, April 20, 2015, thecut.com/2015/04/what-to-do-when-someone-doesnt-get-you.html.
24. Malcolm Gladwell, *Blink: The Power of Thinking Without Thinking* (New York: Back Bay, 2007), 50, Kindle edition.
25. Amy Cuddy, *Presence: Bringing Your Boldest Self to Your Biggest Challenges* (New York: Little, Brown and Company, 2015), 106.
26. Cuddy, *Presence*, 163.
27. Sheryl Sandberg, *Lean In: Women, Work, and the Will to Lead* (New York: Alfred A. Knopf, 2013), 42.
28. Sandberg, *Lean In*, 62.
29. Suzette Caleo and Madeline E. Heilman, "What Could Go Wrong? Some Unintended Consequences of Gender Bias Interventions," *Archives of Scientific Psychology* 7, no. 1 (November 2019): 71–80, dx.doi.org/10.1037/arc0000063.
30. Laura Guillen, "Is the Confidence Gap Between Men and Women a Myth?," *Harvard Business Review*, March 26, 2018, hbr.org/2018/03/is-the-confidence-gap-between-men-and-women-a-myth.
31. Claudia Goldin and Cecilia Rouse, "Orchestrating Impartiality: The Impact of 'Blind' Auditions on Female Musicians," *American Economic Review* 90, no. 4 (September 2000): 738, doi.org/10.1257/aer.90.4.715.
32. Jonatan Pallesen, "Blind Auditions and Gender Discrimination," *jsmp* (blog), May 11, 2019, jsmp.dk/posts/2019–05–12-blindauditions.
33. Andrew Gelman, "Did Blind Orchestra Auditions Really Benefit Women?," *Statistical Modeling, Causal Inference, and Social Science* (blog), May 11, 2019, statmodeling.stat.columbia.edu/2019/05/11/did-blind-orchestra-auditions-really-benefit-women.
34. Jason Dana, "The Utter Uselessness of Job Interviews," *The New York Times*, April 8, 2017, nytimes.com/2017/04/08/opinion/sunday/the-utter-uselessness-of-job-interviews.html.

35. Danielle Kurtzleben, "Why Your Pharmacist Has the Key to Shrinking the Gender Wage Gap," *Vox*, August 19, 2014, vox.com/2014/8/19/6029371/claudia-goldin-gender-wage-gap-temporal-flexibility.

## 4. POSITIVE PSYCHOLOGY GOES TO WAR

1. Daniel Kahneman, *Thinking, Fast and Slow* (New York: Farrar, Straus and Giroux, 2011), 12.
2. The phrase "unskilled intuition" was attributed, in a 2019 edited volume, to Daniel Kahneman and Gary Klein in a 2009 paper, "Conditions for Intuitive Expertise: A Failure to Disagree," *American Psychologist* 64, no. 6 (September 2009): 515–26, doi.org/10.1037/a0016755, though they don't appear to have actually used it there. See Matthew J. Grawitch and David W. Ballard, "Pseudoscience Won't Create a Psychologically Healthy Workplace," in *Creating Psychologically Healthy Workplaces*, ed. Ronald J. Burke and Astrid M. Richardsen (Cheltenham, UK: Edward Elgar, 2019), 44.
3. Gregory Mitchell, "Jumping to Conclusions: Advocacy and Application of Psychological Research," Virginia Public Law and Legal Theory Research Paper, no. 2017–31 (May 2017): 139, ssrn.com/abstract=2973892.
4. Martin E. P. Seligman and Mihaly Csikszentmihalyi, "Positive Psychology: An Introduction," *American Psychologist* 55, no. 1 (January 2000): 5–14, doi.org/10.1037/0003–066X.55.1.5.
5. Rob Hirtz, "Martin Seligman's Journey from Learned Helplessness to Learned Happiness," *Pennsylvania Gazette* 97, no. 3 (January/February 1999), upenn.edu/gazette/0199/hirtz.html.
6. Martin E. P. Seligman, *Helplessness: On Depression, Development, and Death* (New York: W. H. Freeman, 1975).
7. Seligman, interview by Joshua Freedman, November 10, 1999, *EQ Life, Six Seconds*.
8. Seligman's faculty page: ppc.sas.upenn.edu/people/martin-ep-seligman.
9. Barbara Ehrenreich, *Bright-Sided: How the Relentless Promotion of Positive Thinking Has Undermined America* (New York: Metropolitan Books, 2009), 149–50.
10. Daniel Horowitz, *Happier? The History of a Cultural Movement That Aspired to Transform America* (New York: Oxford University Press, 2018), 5.
11. Ehrenreich, *Bright-Sided*, 165.
12. Harris L. Friedman and Nicholas J. L. Brown, "Implications of Debunking the 'Critical Positivity Ratio' for Humanistic Psychology: Introduction to Special Issue," *Journal of Humanistic Psychology* 58, no. 3 (May 2018): 239, doi.org/10.1177/0022167818762227.
13. "Praise for *Positivity*," *Positivity Ratio* (blog), n.d., positivityratio.com/praise.php.
14. Sonja Lyubomirsky, Kennon M. Sheldon, and David Schkade, "Pursuing Happiness: The Architecture of Sustainable Change," *Review of General Psychology* 9, no. 2 (June 2005): 111–31, doi.org/10.1037/1089–2680.9.2.111.
15. Nicholas J. L. Brown and Julia M. Rohrer, "Easy as (Happiness) Pie? A Critical Evaluation of a Popular Model of the Determinants of Well-Being,"

*Journal of Happiness Studies* 21, no. 4 (April 2020): 1285–301, doi.org/10.1007/s10902-019-00128-4.

16. Ruth Whippman, *America the Anxious: How to Calm Down, Stop Worrying, and Find Happiness* (New York: Macmillan, 2016), 195.
17. Todd B. Kashdan and Michael F. Steger, "Challenges, Pitfalls, and Aspirations for Positive Psychology," in *Designing Positive Psychology: Taking Stock and Moving Forward*, ed. Kennon Marshall Sheldon et al. (New York: Oxford University Press, 2011).
18. Horowitz, *Happier?*, 219.
19. Seph Fontane Pennock, "Positive Psychology 1504: Harvard's Groundbreaking Course," *Positive Psychology* (blog), February 18, 2020, positivepsychology.com/harvard-positive-psychology-course-1504/.
20. Martin Seligman and Peter Schulman, *Penn Positive Psychology Center Annual Report*, University of Pennsylvania, May 22, 2018, 1.
21. Martin E. P. Seligman et al., "Positive Education: Positive Psychology and Classroom Interventions," *Oxford Review of Education* 35, no. 3 (June 2009): 300–1, doi.org/10.1080/03054980902934563.
22. Seligman, "Positive Education," 300.
23. Seligman, "Positive Education," 302: "Among students in non-honors classes, the positive psychology programme [that is, Strath Haven] increased Language Arts achievement through eleventh grade." This section doesn't provide any information about the magnitude of this improvement or how many other subgroups' school achievement was measured, meaning it is impossible to evaluate the strength of the result. It stands to reason, though, that if the authors had other positive results to report about grades here, they would have mentioned them.
24. Seligman, "Positive Education," 302.
25. Jane E. Gillham et al., "Preventing Depression in Early Adolescent Girls: The Penn Resiliency and Girls in Transition Programs," in *Handbook of Prevention and Intervention Programs for Adolescent Girls*, ed. Craig W. LeCroy and Joyce Mann (Hoboken, N.J.: John Wiley & Sons, 2008), 128.
26. Gillham, "Preventing Depression in Early Adolescent Girls," 128.
27. Gillham, "Preventing Depression in Early Adolescent Girls," 136.
28. Jesse Singal, "For 80 Years, Young Americans Have Been Getting More Anxious and Depressed, and No One Is Quite Sure Why," *The Cut*, March 13, 2016, thecut.com/2016/03/for-80-years-young-americans-have-been-getting-more-anxious-and-depressed.html.
29. "Penn Resilience Program and Perma™ Workshops," Trustees of the University of Pennsylvania, ppc.sas.upenn.edu/services/penn-resilience-training.
30. "Teaching Resilience, Sense of Purpose in Schools Can Prevent Depression, Anxiety and Improve Grades, According to Research," American Psychological Association, August 2009, apa.org/news/press/releases/2009/08/positive-educate.
31. Steven M. Brunwasser, Jane E. Gillham, and Eric S. Kim, "A Meta-analytic Review of the Penn Resiliency Program's Effect on Depressive

Symptoms," *Journal of Consulting and Clinical Psychology* 77, no. 6 (December 2009): 1042–54, doi:10.1037/a0017671.

32. Anastasios Bastounis et al., "The Effectiveness of the Penn Resiliency Program (PRP) and Its Adapted Versions in Reducing Depression and Anxiety and Improving Explanatory Style: A Systematic Review and Meta-Analysis," *Journal of Adolescence* 52 (October 2016): 37, doi.org/10.1016/j.adolescence.2016.07.004.

33. Sarah E. Hetrick, Georgina R. Cox, and Sally N. Merry, "Where to Go from Here? An Exploratory Meta-analysis of the Most Promising Approaches to Depression Prevention Programs for Children and Adolescents," *International Journal of Environmental Research and Public Health* 12, no. 5 (May 2015): 4758–95, doi:10.3390/ijerph120504758. This meta-analysis, which deals with both PRP and other programs geared at preventing depression in children and adolescents, showed that PRP was associated with "small positive effects in reducing depressive symptoms in youth at post intervention and up to 1-year follow-up, compared with those who received no intervention." But, as the authors note, many of the PRP variants they studied were targeted rather than universal—that is, the study participants already had signs of mental-health problems and thus more room to improve than healthy populations, meaning larger results could be expected. Sure enough, the authors found that targeted depression-prevention programs (the PRP ones and others) were associated with almost twice as large a reduction in depression symptoms, a year out, as the universal ones.

    Moreover, in many of the studies evaluated in this meta-analysis, the facilitators were mental-health experts rather than trained laypeople (such as teachers). In fact, when the researchers grouped together all the studies in which the interventions being evaluated were delivered by non-mental-health professionals, a year later there was no statistically significant improvement in the mental health of the participants.

    The version of PRP Seligman sells via the Positive Psychology Center derives great currency from the idea that it can be delivered to universal populations of students or soldiers or professionals by laypeople who can be quickly trained as PRP facilitators. This meta-analysis does not offer much evidence for the program's effectiveness under these circumstances. That said, because of the way the results are presented—particularly the lumping together of PRP and non-PRP interventions—certain key questions remain unanswered, and the authors didn't respond to multiple emails asking them to clarify certain points.

34. Ethan Watters, "We Aren't the World," *Pacific Standard*, February 25, 2013, psmag.com/social-justice/joe-henrich-weird-ultimatum-game-shaking-up-psychology-economics-53135.

35. Seligman and Schulman, *Penn Positive Psychology Center Annual Report*, 1.

36. David S. Cloud, "Military Eases Its Rules for Mobilizing Reserves," *The New York Times*, January 12, 2007, nytimes.com/2007/01/12/washington/12guard.html.

37. See, for example, Mark Wasson, "One Weekend a Month, Two Weeks a Year Is a Lie," *U.S. Patriot* (blog), n.d., blog.uspatriottactical.com/one-weekend-a-month-two-weeks-a-year-is-a-lie.
38. U.S. Department of Veterans Affairs, *Health Study for a New Generation of U.S. Veterans* (Washington, D.C., 2010), publichealth.va.gov/epidemiology/studies/new-generation/index.asp.
39. Jaimie L. Gradus, "PTSD: National Center for PTSD," U.S. Department of Veterans Affairs, October 14, 2019, ptsd.va.gov/professional/treat/essentials/epidemiology.asp#:~:text=U.S.%20National%20Comorbidity%20Survey%20Replication&text=The%20NCS%2DR%20estimated%20the,be%206.8%25%20(1.
40. "What is Posttraumatic Stress Disorder?," *American Psychiatric Association*, n.d., psychiatry.org/patients-families/ptsd/what-is-ptsd.
41. Kander, interview by Hayes, NBC News, September 11, 2019.
42. Deborah Sontag and Lizette Alvarez, "Iraq Veterans Leave a Trail of Death and Heartbreak in U.S.," *The New York Times*, January 13, 2008, nytimes.com/2008/01/13/world/americas/13iht-vets.1.9171147.html.
43. Veterans Administration, "Military Sexual Trauma," 2015, www.mentalhealth.va.gov/docs/mst_general_factsheet.pdf.
44. Mark Benjamin, "'War on Terror' Psychologist Gets Giant No-Bid Contract," *Salon*, October 14, 2010, salon.com/2010/10/14/army_contract_seligman/.
45. Martin E. P. Seligman, *Flourish: A Visionary New Understanding of Happiness and Well-Being* (New York: Free Press, 2011), 163.
46. Department of Defense Appropriations for Fiscal Year 2011: Subcommittee of the Committee on Appropriations, 112th Cong. (2010) (statement of John M. McHugh, Secretary of the U.S. Army, and George W. Casey, Chief of Staff of the U.S. Army), gpo.gov/fdsys/pkg/CHRG-111shrg89104380/html/CHRG-111shrg89104380.htm.
47. Karen J. Reivich, Martin E. P. Seligman, and Sharon McBride, "Master Resilience Training in the U.S. Army," *American Psychologist* 66, no. 1 (2011): 25–34, doi.org/10.1037/a0021897.
48. Nicholas J. L. Brown, "A Critical Examination of the U.S. Army's Comprehensive Soldier Fitness Program," *Winnower* (July 2015), doi:10.15200/winn.143751.17496.
49. Lisa M. Najavits, "The Problem of Dropout from 'Gold Standard' PTSD Therapies," *F1000Prime Reports* 7, no. 43 (April 2015): doi.org/10.12703/p7–43.
50. Emily M. Johnson et al., "Veterans' Thoughts About PTSD Treatment and Subsequent Treatment Utilization," *International Journal of Cognitive Therapy* 10, no. 2 (June 2017): 138–60, doi.org/10.1521/ijct_2017_10_02.
51. Heidi M. Zinzow et al., "Connecting Active Duty and Returning Veterans to Mental Health Treatment: Interventions and Treatment Adaptations that May Reduce Barriers to Care," *Clinical Psychology Review* 32, no. 8 (2012): 741–53, doi: 10.1016/j.cpr.2012.09.002.
52. Charles Tilly, *Why? What Happens When People Give Reasons . . . and Why* (Princeton, N.J.: Princeton University Press, 2006), 60.

53. Daniel DeFraia, "The Unknown Legacy of Military Mental Health Programs," *The War Horse*, October 2, 2019, thewarhorse.org/newsroom-archive/the-unknown-legacy-of-military-mental-health-programs.
54. Seligman, *Flourish*, 163.
55. David Vergun, "Study Concludes Master Resilience Training Effective," U.S. Army, January 25, 2012, army.mil/article/72431/Study_concludes_Master_Resilience_Training_effective.
56. Roy Eidelson and Stephen Soldz, "Does Comprehensive Soldier Fitness Work? CSF Research Fails the Test," Coalition for an Ethical Psychology, Working Paper no. 1 (2012), royeidelson.com/wp-content/uploads/2017/05/Eidelson-Soldz-CSF_Research_Fails_the_Test.pdf.
57. Amy Novotney, "Strong in Mind and Body," *Monitor on Psychology* 40, no. 11 (December 2009): 40, web.archive.org/web/20180306055502/http://apa.org/monitor/2009/12/army-program.aspx.
58. Eidelson and Soldz, "Does Comprehensive Soldier Fitness Work?," 1.
59. Colonel Richard Franklin Timmons II, "The United States Army Comprehensive Soldier Fitness: A Critical Look" (master's thesis, United States Army War College, 2013), apps.dtic.mil/dtic/tr/fulltext/u2/a590241.pdf.
60. Institute of Medicine, *Preventing Psychological Disorders in Service Members and Their Families: An Assessment of Programs* (Washington, D.C.: National Academies Press, 2014), doi.org/10.17226/18597.
61. Gregg Zoroya, "Army Morale Low Despite 6-Year, $287M Optimism Program," *USA Today*, April 16, 2015, usatoday.com/story/news/nation/2015/04/16/army-survey-morale/24897455/.
62. "Comprehensive Airman Fitness on the Go," press release, July 16, 2015, af.mil/News/Article-Display/Article/608910/comprehensive-airman-fitness-on-the-go/.
63. Dick Donovan, "The Tragic Life of an American Hero," *Weekly World News*, July 9, 1985.

### 5. WHO HAS GRIT?

1. Angela Lee Duckworth, "Grit: The Power of Passion and Perseverance," *TED*, April 2013, ted.com/talks/angela_lee_duckworth_grit_the_power_of_passion_and_perseverance.
2. Julie Scelfo, "Angela Duckworth on Passion, Grit and Success," *The New York Times*, April 8, 2016, nytimes.com/2016/04/10/education/edlife/passion-grit-success.html.
3. Anya Kamenetz, "MacArthur 'Genius' Angela Duckworth Responds to a New Critique of Grit," *All Things Considered*, NPR, May 25, 2016, npr.org/sections/ed/2016/05/25/479172868/angela-duckworth-responds-to-a-new-critique-of-grit.
4. "Grit Scale," *Angela Duckworth* (blog), n.d., angeladuckworth.com/grit-scale.
5. Angela Lee Duckworth et al., "Deliberate Practice Spells Success: Why Grittier Competitors Triumph at the National Spelling Bee," *Social Psychological and Personality Science* 2, no. 2 (March 2011): 174–81, doi:10.1177/1948550610385872.

6. Lauren Eskreis-Winkler et al., "The Grit Effect: Predicting Retention in the Military, the Workplace, School and Marriage," *Frontiers in Psychology* 5, no. 36 (2014): doi.org/10.3389/fpsyg.2014.00036.
7. Kate Zernike, "Testing for Joy and Grit? Schools Nationwide Push to Measure Students' Emotional Skills," *The New York Times*, February 29, 2016, nytimes.com/2016/03/01/us/testing-for-joy-and-grit-schools-nationwide-push-to-measure-students-emotional-skills.html.
8. Tovia Smith, "Does Teaching Kids to Get 'Gritty' Help Them Get Ahead?," *Morning Edition*, NPR, March 17, 2014, npr.org/sections/ed/2014/03/17/290089998/does-teaching-kids-to-get-gritty-help-them-get-ahead.
9. Loretta Kalb, "Grit and Gratitude Join Reading, Writing, and Arithmetic on Report Cards," *Sacramento Bee*, January 26, 2015, sacbee.com/news/local/education/article8217330.html.
10. "Character," KIPP, n.d., kipp.org/approach/character.
11. Martin E. P. Seligman, *Flourish: A Visionary New Understanding of Happiness and Well-Being* (New York: Free Press, 2011), 103.
12. Ethan W. Ris, "Grit: A Short History of a Useful Concept," *Journal of Educational Controversy* 10, no. 1 (2015), cedar.wwu.edu/jec/vol10/iss1/3.
13. Ris, "Grit," 3.
14. Ris, "Grit," 4.
15. Ris, "Grit," 5.
16. Angela Duckworth, *Grit: The Power of Passion and Perseverance* (New York: Scribner, 2016), 10.
17. Tena Vukasović and Denis Bratko, "Heritability of Personality: A Meta-analysis of Behavior Genetic Studies," *Psychological Bulletin* 141, no. 4 (July 2015): 769–85, doi.org/10.1037/bul0000017.
18. Angela L. Duckworth et al., "Grit: Perseverance and Passion for Long-Term Goals," *Journal of Personality and Social Psychology* 92, no. 6 (2007): 1087–101, doi.org/10.1037/0022–3514.92.6.1087.
19. Angela Lee Duckworth, "Grit: The Power of Passion and Perseverance," April 2013, TED video, ted.com/talks/angela_lee_duckworth_grit_the_power_of_passion_and_perseverance.
20. Angela Lee Duckworth and Patrick D. Quinn, "Development and Validation of the Short Grit Scale (Grit-S)," *Journal of Personality Assessment* 91, no. 2 (February 2009): 166–74, doi:10.1080/00223890802634290.
21. Angela Duckworth, Patrick D. Quinn, and Martin E. P. Seligman, "Positive Predictors of Teacher Effectiveness," *Journal of Positive Psychology* 4, no. 6 (November 2009): 540–47, doi:10.1080/17439760903157232.
22. The 2014 study in question is Claire Robertson-Kraft and Angela Lee Duckworth, "True Grit: Trait-Level Perseverance and Passion for Long-Term Goals Predicts Effectiveness and Retention Among Novice Teachers," *Teachers College Record* 116, no. 3 (2014). It didn't even use the grit scale. Instead, the authors used a completely different instrument—a scale the College Board developed in 1985 that evaluates students' extracurricular activities, giving them points based on both how long they participated in the activities and their level of achievement in them. The five-point scale gives one point for multiyear participation in an activity

and two points for "high achievement," such as team captainship, in that activity. This instrument had already been shown, by the College Board, to be an effective predictor of college success.

On its face, though, this scale seems to violate the spirit of grit in a rather fundamental way: if two high school swimmers practice the exact same amount, both swim varsity for two years, but the more naturally athletic one wins a state title while the less gifted one does not, the champion would get three points for the swimming entry, while the mediocre swimmer would get one. This clearly would not reflect a difference in "grit," which is meant to describe effort and consistency of interest, not achievement.

Despite these rather profound differences Duckworth and Robertson-Kraft simply assumed the College Board measure to be a proxy for grit and adapted it as is (other than a trivial change to the scoring system). They had research assistants use it to assign scores to a cohort of novice Teach for America teachers, on the basis of the college activities listed on their résumés, and found that this measure predicted teacher retention and quality (as measured by TFA) better than the teachers' college GPA or SAT scores. But this really isn't a study of grit, and this study, like some of Duckworth's other most famous ones, also didn't control for conscientiousness. (To be clear, if the researchers showed that the original grit scale and the College Board method were highly correlated, that would attenuate most of these criticisms, but they didn't even check.)

23. Eskreis-Winkler, "The Grit Effect."
24. Marcus Crede, Michael C. Tynan, and Peter D. Harms, "Much Ado About Grit: A Meta-Analytic Synthesis of the Grit Literature," *Journal of Personality and Social Psychology* 113, no. 3 (2017): 492–511, doi.org/10.1037/pspp0000102.
25. Crede, "Much Ado About Grit," 495.
26. Crede, "Much Ado About Grit," 495.
27. Kamenetz, "MacArthur 'Genius' Angela Duckworth Responds to a New Critique of Grit."
28. Arthur E. Poropat, "A Meta-analysis of the Five-Factor Model of Personality and Academic Performance," *Psychological Bulletin* 135, no. 2 (2009): 322–38, doi.org/10.1037/a0014996.
29. Barbara Dumfart and Aljoscha C. Neubauer, "Conscientiousness Is the Most Powerful Noncognitive Predictor of School Achievement in Adolescents," *Journal of Individual Differences* 37, no. 1 (February 2016): 8–15, doi:10.1027/1614-0001/a000182. And a good layperson-friendly rundown is here: Brian Resnick, "IQ, Explained in 9 Charts," *Vox*, May 24, 2016, vox.com/2016/5/24/11723182/iq-test-intelligence.
30. Ian J. Deary et al., "Intelligence and Educational Achievement," *Intelligence* 35, no. 1 (February 2007): 13–21, doi.org/10.1016/j.intell.2006.02.001.
31. Duckworth et al., "Grit: Perseverance and Passion for Long-Term Goals."
32. Brent W. Roberts et al., "A Systematic Review of Personality Trait Change Through Intervention," *Psychological Bulletin* 143, no. 2 (January 2017): 117–41, doi.org/10.1037/bul0000088.

33. Sue Martin, Lindsay G. Oades, and Peter Caputi, "A Step-Wise Process of Intentional Personality Change Coaching," *International Coaching Psychology Review*, 9, no. 2 (2014): 181–95; Sue Martin, Lindsay G. Oades, and Peter Caputi, "Intentional Personality Change Coaching: A Randomised Controlled Trial of Participant Selected Personality Facet Change Using the Five-Factor Model of Personality," *International Coaching Psychology Review* 9, no. 2 (2014): 196–209.
34. Daniel Engber, "Is 'Grit' Really the Key to Success?," *Slate*, May 8, 2016, slate.com/articles/health_and_science/cover_story/2016/05/angela_duckworth_says_grit_is_the_key_to_success_in_work_and_life_is_this.html.
35. David J. Disabato, Fallon R. Goodman, and Todd B. Kashdan, "Is Grit Relevant to Well-Being and Strengths? Evidence Across the Globe for Separating Perseverance of Effort and Consistency of Interests," *Journal of Personality* 87, no. 2 (April 2018): 1, doi.org/10.1111/jopy.12382.
36. Chen Zissman and Yoav Ganzach, "In a Representative Sample Grit Has a Negligible Effect on Educational and Economic Success Compared to Intelligence," *Social Psychological and Personality Science* (original manuscript, 2020): 1–8, doi.org/10.1177/1948550620920531.
37. Angela Duckworth, "Don't Grade Schools on Grit," *The New York Times*, March 26, 2016, nytimes.com/2016/03/27/opinion/sunday/dont-grade-schools-on-grit.html.
38. Melissa Dahl, "Don't Believe the Hype About Grit, Pleads the Scientist Behind the Concept," *The Cut*, May 9, 2016, thecut.com/2016/05/dont-believe-the-hype-about-grit-pleads-the-scientist-behind-the-concept.html.
39. Crede, "Much Ado About Grit," 503.
40. Thomas L. Dynneson, *City-State Civism in Ancient Athens: Its Real and Ideal Expressions* (New York: Peter Lang, 2008), 1.
41. David F. Labaree, *Someone Has to Fail: The Zero-Sum Game of Public Schooling* (Cambridge, Mass.: Harvard University Press, 2012), 19–20.
42. David Casalaspi, "The Day Education Failed," *Green & Write*, November 11, 2016, education.msu.edu/green-and-write/2016/the-day-education-failed.
43. Associated Press, "Urges Return to Schools of McGuffey's Readers," *The New York Times*, May 9, 1935, timesmachine.nytimes.com/timesmachine/1935/05/09/94604752.html?auth=login-email&pageNumber=19.
44. Charles E. Greenawalt, II, *Character Education in America* (Harrisburg, Penn.: Commonwealth Foundation for Public Policy Alternatives, 1996), 4, files.eric.ed.gov/fulltext/ED398327.pdf.
45. "Education Chief Lauds Religion at School Parley," *Los Angeles Times*, November 9, 1985, articles.latimes.com/1985–11–09/local/me-3591_1_bennett.
46. Michael Watz, "An Historical Analysis of Character Education," *Journal of Inquiry and Action in Education* 4, no. 2 (2011): 34–53, files.eric.ed.gov/fulltext/EJ1134548.pdf.
47. President William Jefferson Clinton, "State of the Union Address," Washington, D.C., January 23, 1996, clintonwhitehouse2.archives.gov/WH/New/other/sotu.html.

48. Associated Press, "President Clinton's Message to Congress on the State of the Union," *The New York Times*, February 5, 1997, nytimes.com/1997/02/05/us/president-clinton-s-message-to-congress-on-the-state-of-the-union.html.
49. James S. Leming, "Whither Goes Character Education? Objectives, Pedagogy, and Research in Education Programs," *Journal of Education* 179, no. 2 (1997): 17, eric.ed.gov/?id=EJ562093.
50. Leming, "Whither Goes Character Education?," 17.
51. Leming, "Whither Goes Character Education?," 17, 28.
52. Paul Tough, *How Children Succeed: Grit, Curiosity, and the Hidden Power of Character* (Boston: Mariner Books, 2012), 60.
53. *Efficacy of Schoolwide Programs to Promote Social and Character Development and Reduce Problem Behavior in Elementary School Children*, prepared by the Social and Character Development Research Consortium in cooperation with the U.S. Department of Education (Washington, D.C., 2010), ies.ed.gov/ncer/pubs/20112001/pdf/20112001a.pdf.
54. Michael B. Katz, "Public Education as Welfare," *Dissent*, Summer 2010, dissentmagazine.org/article/public-education-as-welfare.
55. Paul Tough, "How Kids Learn Resilience," *The Atlantic*, June 2016, theatlantic.com/magazine/archive/2016/06/how-kids-really-succeed/480744.
56. Corey DeAngelis, Heidi Holmes Erickson, and Gary Ritter, "Is Pre-kindergarten an Educational Panacea? A Systematic Review and Meta-analysis of Scaled-Up Pre-kindergarten in the United States," Working Paper 2017–08, University of Arkansas Department of Education Reform Research Paper Series, February 16, 2017, papers.ssrn.com/sol3/papers.cfm?abstract_id=2920635. One disappointing result, though, was a study of children in Tennessee: Mark W. Lipsey, Dale C. Farran, and Kelley Durkin, "Effects of the Tennessee Prekindergarten Program on Children's Achievement and Behavior Through Third Grade," *Early Child Research Quarterly* 45, no. 4 (2018): 155–76, sciencedirect.com/science/article/pii/S0885200618300279; Charles Murray, "Response: Weighing the Evidence," *Boston Review*, September/October 2012, bostonreview.net/archives/BR37.5/ndf_charles_murray_social_mobility.php.
57. As of 2013, according to a UN report, almost a quarter of American children lived in families pulling in less than 50 percent of the median income—the second highest such figure in the entire OECD cohort of developed nations, only ahead of Romania. The United States also compares poorly with other wealthy, industrialized nations when it comes to food insecurity, which is a proxy for all sorts of health and behavioral problems. According to a UNICEF study, about 19.6 percent of American households with children have moderate or severe food insecurity, compared with about 12 percent in Canada, 6 percent in France, and 4.9 percent in Germany (the UK is an outlier here, scoring slightly worse than the United States). These sorts of cross-national comparisons can be fraught for various reasons, but there's a general consensus among researchers that the United States does a poorer job than other countries in its economic echelon of helping shield children from the most damaging effects of poverty and neglect.

58. David F. Labaree, *Someone Has to Fail: The Zero-Sum Game of Public Schooling* (Cambridge, Mass.: Harvard University Press, 2010), 171.
59. Duckworth, *Grit*, 187.
60. Linda F. Nathan, *When Grit Isn't Enough: A High School Principal Examines How Poverty and Inequality Thwart the College-for-All Promise* (Boston: Beacon Press, 2017).
61. Nathan, *When Grit Isn't Enough*, 4.
62. Nathan, *When Grit Isn't Enough*, 20–21.
63. Nathan, *When Grit Isn't Enough*, 76.
64. Nathan, *When Grit Isn't Enough*, 23.
65. Duckworth, *Grit*, 408–409, Kindle edition.
66. Tough, *How Children Succeed*, 194.
67. Ris, "Grit," 7–8.
68. Darrin Donnelly, *Old School Grit: Times May Change, But the Rules for Success Never Do, Sports for the Soul*, vol. 2 (Lenexa, Kans.: Shamrock New Media, 2016).
69. Jung Choi et al., *Millennial Homeownership: Why Is It So Low, and How Can We Increase It?* (Washington, D.C.: Urban Institute, 2018), v.
70. Lucinda Shen, "Millennials Are Worth Half as Much as Their Parents Were at the Same Age," *Fortune*, January 13, 2017, fortune.com/2017/01/13/millennial-boomer-worth-income-study.

## 6. THE BIAS TEST

1. Matt Stevens, "Starbucks C.E.O. Apologizes After Arrests of 2 Black Men," *The New York Times*, April 15, 2018, nytimes.com/2018/04/15/us/starbucks-philadelphia-black-men-arrest.html; Elisha Fieldstadt, "White Starbucks Manager Claims Racial Bias in Her Firing After Arrests of 2 Black Men," *NBC News*, October 21, 2019, nbcnews.com/news/us-news/white-starbucks-manager-claims-racial-bias-her-firing-after-arrests-n1074431.
2. Joel Schwarz, "Roots of Unconscious Prejudice Affect 90 to 95 Percent of People, Psychologists Demonstrate at Press Conference," *University of Washington News*, September 29, 1998, washington.edu/news/1998/09/29/roots-of-unconscious-prejudice-affect-90-to-95-percent-of-people-psychologists-demonstrate-at-press-conference.
3. Jill D. Kester, "A Revolution in Social Psychology," *APS Observer* 14, no. 6 (July/August 2001), psychologicalscience.org/observer/0701/family.html.
4. Malcolm Gladwell, *Blink: The Power of Thinking Without Thinking* (New York: Back Bay, 2007), 85, Kindle edition.
5. Nicholas Kristof, "Our Biased Brains," *The New York Times*, May 7, 2015, nytimes.com/2015/05/07/opinion/nicholas-kristof-our-biased-brains.html.
6. Mahzarin R. Banaji and Anthony G. Greenwald, *Blindspot: Hidden Biases of Good People* (New York: Delacorte Press, 2013), 47.
7. NBC, "Pride and Prejudice: Prejudice in America," *Dateline*, March 19, 2000.

8. Banaji, *Blindspot*, 209.
9. B. A. Nosek, M. R. Banaji, and A. G. Greenwald, "Harvesting Implicit Group Attitudes and Beliefs from a Demonstration Web Site," *Group Dynamics: Theory, Research, and Practice* 6, no. 1 (2002): 101–15, psycnet.apa.org/record/2002-10827-009.
10. Shankar Vedantam, "See No Bias," *The Washington Post*, January 23, 2005, washingtonpost.com/archive/lifestyle/magazine/2005/01/23/see-no-bias/a548dee4-4047-4397-a253-f7f780fae575.
11. Banaji and Greenwald, *Blindspot*, 45.
12. John Tierney, "A Shocking Test of Bias," *The New York Times*, November 18, 2008, tierneylab.blogs.nytimes.com/2008/11/18/a-shocking-test-of-bias.
13. Gladwell, *Blink*, 84.
14. April Dembosky, "UCSF Doctors, Students Confront Their Own Unconscious Bias," KQED, August 4, 2015, kqed.org/stateofhealth/56311/ucsf-doctors-students-confront-their-own-unconscious-bias.
15. "Exposing Bias at Google," *The New York Times*, September 24, 2014.
16. "Facebook's Gender Bias Goes So Deep It's in the Code," *ThinkProgress*, May 2, 2017.
17. "Clients and Testimonials," Fair and Impartial Policing, fipolicing.com/clients-and-testimonials.
18. Kristin A. Lane et al., "Understanding and Using the Implicit Association Test: IV," in *Implicit Measures of Attitudes*, ed. Bernd Wittenbrink and Norbert Schwarz (New York: Guilford Press, 2007), 59–102, faculty.washington.edu/agg/pdf/Lane%20et%20al.UUIAT4.2007.pdf. This chapter includes a table running down the test-retest reliabilities for the race IAT that had been published to that point: $r = 0.32$ in a study consisting of four race IAT sessions conducted with two weeks between each; $r = 0.65$ in a study in which two tests were conducted twenty-four hours apart; and $r = 0.39$ in a study in which the two tests were conducted during the same session (but in which one used names and the other used pictures). In 2014, using a large sample, the researcher Yoav Bar-Anan and Nosek reported a race IAT test-retest reliability of $r = 0.4$.
19. C. K. Lai and M. E. Wilson, "Measuring Implicit Intergroup Biases," unpublished manuscript (2020); Anthony G. Greenwald and Calvin K. Lai, "Implicit Social Cognition," *Annual Review of Psychology* 71, no. 1 (2020): 419–45, doi.org/10.1146/annurev-psych-010419-050837.
20. Mahzarin R. Banaji and Anthony G. Greenwald, *Blindspot: Hidden Biases of Good People* (New York: Delacorte Press, 2013).
21. Allan W. Wicker, "Attitudes Versus Actions: The Relationship of Verbal and Overt Behavioral Responses to Attitude Objects," *Journal of Social Issues* 25, no. 4 (October 1969): 41–78, doi.org/10.1111/j.1540-4560.1969.tb00619.x.
22. Anthony G. Greenwald et al., "Understanding and Using the Implicit Association Test: III. Meta-analysis of Predictive Validity," *Journal of Personality and Social Psychology* 97, no. 1 (2009): 17–41.
23. Frederick L. Oswald et al., "Predicting Ethnic and Racial Discrimination: A Meta-analysis of IAT Criterion Studies," *Journal of Personality and Social Psychology* 105, no. 2 (2013): 171–92.

24. Anthony G. Greenwald, Mahzarin Banaji, and Brian Nosek, "Statistically Small Effects of the Implicit Association Test Can Have Societally Large Effects," *Journal of Personality and Social Psychology* 108, no. 4 (2015): 553–61.
25. Part of the problem is that the extant studies are seriously statistically underpowered. Or at least that was the finding of the researchers Rickard Carlsson and Jens Agerström, who, in a meta-analysis of their own published in the *Scandinavian Journal of Psychology* in 2016, argued that "attempting to meta-analytically test the correlation between IAT and discrimination thus appears futile. We are, essentially, chasing noise, and simply cannot expect any strong, or even moderate, correlations, based on the current literature."
26. Patrick S. Forscher et al., "A Meta-analysis of Procedures to Change Implicit Measures," *Journal of Personality and Social Psychology* 117, no. 3 (2019): 522–59, psycnet.apa.org/record/2019-31306-001.
27. Olivia Goldhill, "The World Is Relying on a Flawed Psychological Test to Fight Racism," *Quartz*, December 3, 2017, qz.com/1144504/the-world-is-relying-on-a-flawed-psychological-test-to-fight-racism.
28. Chloë FitzGerald et al., "Interventions Designed to Reduce Implicit Prejudices and Implicit Stereotypes in Real World Contexts: A Systematic Review," *BMC Psychology* 7, no. 29 (2019), doi.org/10.1186/s40359-019-0299-7.
29. Forscher, "Meta-analysis of Procedures to Change Implicit Measures."
30. Hal R. Arkes and Philip E. Tetlock, "Attributions of Implicit Prejudice, or 'Would Jesse Jackson "Fail" the Implicit Association Test?,'" *Psychological Inquiry* 15, no. 4 (2004): 257–78.
31. Eric Luis Uhlmann, Victoria L. Brescoll, and Elizabeth Levy Paluck, "Are Members of Low Status Groups Perceived as Bad, or Badly Off? Egalitarian Negative Associations and Automatic Prejudice," *Journal of Experimental Social Psychology* 42, no. 4 (2006): 491–99.
32. Cynthia M. Frantz et al., "A Threat in the Computer: The Race Implicit Association Test as a Stereotype Threat Experience," *Personality and Social Psychology Bulletin* 30, no. 12 (2004): 1611–24.
33. Sam G. McFarland and Zachary Crouch, "A Cognitive Skill Confound on the Implicit Association Test," *Social Cognition* 20, no. 6 (2002): 483–510.
34. Hart Blanton and Elif G. Ikizer, "The Bullet-Point Bias: How Diluted Science Communications Can Impede Social Progress," in *Politics of Social Psychology*, ed. Jarret T. Crawford and Lee Jussim (New York: Routledge, 2018).
35. Eduardo Bonilla-Silva, "Rethinking Racism: Toward a Structural Interpretation," *American Sociological Review* 62, no. 3 (1997): 467, doi.org/10.2307/2657316.
36. William B. Johnston et al., *Workforce 2000: Work and Workers for the 21st Century* (Indianapolis: Hudson Institute, 1987), 14, files.eric.ed.gov/fulltext/ED290887.pdf.
37. Rohini Anand and Mary-Frances Winters, "A Retrospective View of Corporate Diversity Training from 1964 to the Present," *Academy of Management Learning and Education* 7, no. 3 (2008): 359.
38. George B. Leonard, *Education and Ecstasy* (New York: Delta, 1968), 197–98.

39. Lisette Voytko, "74% of Americans Support George Floyd Protests, Majority Disapprove of Trump's Handling, Poll Says," *Forbes*, June 9, 2020, forbes.com/sites/lisettevoytko/2020/06/09/74-of-americans-support-george-floyd-protests-majority-disapprove-of-trumps-handling.
40. Elisabeth Lasch-Quinn, *Race Experts: How Racial Etiquette, Sensitivity Training, and New Age Therapy Hijacked the Civil Rights Revolution* (Lanham, Md.: Rowman and Littlefield, 2001), 7.
41. Lasch-Quinn, *Race Experts*, 125.
42. Frantz Fanon, *Toward the African Revolution: Political Essays*, trans. Haakon Chevalier (New York: Grove Press, 1964), 34.
43. Robin DiAngelo, *White Fragility: Why It's So Hard for White People to Talk About Racism* (Boston: Beacon, 2018), 33.
44. DiAngelo, *White Fragility*, 132.
45. DiAngelo, *White Fragility*, 136.
46. DiAngelo, *White Fragility*, 82.
47. Google Dictionary, s.v. "Microaggression."
48. Scott O. Lilienfeld, "Microaggressions: Strong Claims, Inadequate Evidence," *Perspectives on Psychological Science* 12, no. 1 (January 2017): 138–69, doi.org/10.1177/1745691616659391.
49. Emily Ekins, *The State of Free Speech and Tolerance in America* (Washington, D.C.: Cato Institute, 2017).
50. Susan Edelman, Selim Algar, and Aaron Feis, "Richard Carranza Held 'White-Supremacy Culture' Training for School Admins," *New York Post*, May 20, 2019, nypost.com/2019/05/20/richard-carranza-held-doe-white-supremacy-culture-training.
51. John McWhorter, "Antiracism, Our Flawed New Religion," *The Daily Beast*, July 27, 2015, thedailybeast.com/antiracism-our-flawed-new-religion.
52. Jonathan Kahn, *Race on the Brain: What Implicit Bias Gets Wrong About the Struggle for Racial Justice* (New York: Columbia University Press, 2017), 153.
53. Jesse Singal, "Psychology's Favorite Tool for Measuring Racism Isn't Up to the Job," *The Cut*, January 11, 2017, thecut.com/2017/01/psychologys-racism-measuring-tool-isnt-up-to-the-job.html. (Originally published on the Science of Us vertical, which was later folded into *The Cut*.)
54. Banaji and Greenwald, *Blindspot*, 209.
55. Mahzarin R. Banaji and Anthony G. Greenwald, "Appendix 1: Are Americans Racist?," in *Blindspot: Hidden Biases of Good People*, 2nd ed. (New York: Bantam Books, 2016), 184.
56. Kriston McIntosh et al., "Examining the Black-White Wealth Gap," Brookings, February 27, 2020, brookings.edu/blog/up-front/2020/02/27/examining-the-black-white-wealth-gap.
57. *Report to the United Nations on Racial Disparities in the U.S. Criminal Justice System* (Washington, D.C.: Sentencing Project, 2018), 1.
58. Marianne Bertrand and Sendhil Mullainathan, "Are Emily and Greg More Employable than Lakisha and Jamal? A Field Experiment on Labor Market Discrimination," *American Economic Review* 94, no. 4 (September 2004): 991–1013, doi:10.1257/0002828042002561.

59. Michael Selmi, "The Paradox of Implicit Bias and a Plea for a New Narrative" (George Washington University Law School Public Law Research Paper, 2017), 3–4, ssrn.com/abstract=3026381.
60. Nicolas Jacquemet and Constantine Yannelis, "Indiscriminate Discrimination: A Correspondence Test for Ethnic Homophily in the Chicago Labor Market," *Labour Economics* 19, no. 6 (December 2012): 824–32, doi .org/10.1016/j.labeco.2012.08.004.
61. David J. Deming et al., "The Value of Postsecondary Credentials in the Labor Market: An Experimental Study," *American Economic Review* 106, no. 3 (March 2016): 778–806, doi:10.1257/aer.20141757.
62. Uri Simonsohn, "Greg vs. Jamal: Why Didn't Bertrand and Mullainathan (2004) Replicate?," *Data Colada* (blog), September 6, 2016, datacolada.org/51.
63. Lincoln Quillian et al., "Meta-analysis of Field Experiments Shows No Change in Racial Discrimination in Hiring over Time," *Proceedings of the National Academy of Sciences of the United States of America* 114, no. 41 (October 2017): 10870–75, doi.org/10.1073/pnas.1706255114.
64. Office of Public Affairs, U.S. Department of Justice, "Justice Department Announces Findings of Two Civil Rights Investigations in Ferguson, Missouri," press release, March 4, 2015, emphasis added, justice.gov/opa/pr /justice-department-announces-findings-two-civil-rights-investigations -ferguson-missouri.
65. U.S. Department of Justice, Civil Rights Division, and U.S. Attorney's Office, Northern District of Illinois, *Investigation of the Chicago Police Department* (Washington, D.C., 2017), 144, justice.gov/opa/file/925846 /download.
66. Ann Choi, Keith Herbert, and Olivia Winslow, "Long Island Divided," *Newsday*, November 17, 2019.
67. Margery Austin Turner et al., *Housing Discrimination Against Racial and Ethnic Minorities 2012*, prepared by the Urban Institute in cooperation with the U.S. Department of Housing and Urban Development (Washington, D.C., 2013), huduser.gov/portal/Publications/pdf/HUD-514 _HDS2012.pdf.
68. Jennifer Lynn Eberhardt, *Biased: Uncovering the Hidden Prejudice That Shapes What We See, Think, and Do* (New York: Viking, 2019), 80.
69. New York State Assembly Speaker Carl E. Heastie, "Assembly Passes Eric Garner Anti-Chokehold Act," press release, June 8, 2020, nyassembly.gov /Press/files/20200608a.php.
70. Kahn, *Race on the Brain*, 141; Jim Murdoch and Ralph Roche, "The European Convention on Human Rights and Policing: A Handbook for Police Officers and Other Law Enforcement Officials," *Council of Europe* (December 2013), 24.
71. Conrad G. Brunk, "Public Knowledge, Public Trust: Understanding the 'Knowledge Deficit,'" *Community Genetics* 9, no. 3 (2006): 178–83, doi:10.1159/000092654.

7. NON-REPLICABLE

1. Neuroskeptic, "Social Priming—Does It Work After All?," *Discover*, October 13, 2016, discovermagazine.com/mind/social-priming-does-it-work-after-all.
2. "Republic," in *A Plato Reader: Eight Essential Dialogues*, ed. C. D. C. Reeve (Indianapolis: Hackett, 2012), 541–2.
3. Anthony Pratkanis, "The Cargo-Cult Science of Subliminal Persuasion," *Skeptical Inquirer* 16, no. 3 (Spring 1992): 260–72, skepticalinquirer.org/1992/04/the-cargo-cult-science-of-subliminal-persuasion.
4. Norman Cousins, "Smudging the Subconscious," *Saturday Review*, October 5, 1957, 20.
5. Pratkanis, "The Cargo-Cult Science of Subliminal Persuasion," 261.
6. Maarten Derksen, *Histories of Human Engineering: Tact and Technology* (Cambridge, UK: Cambridge University Press, 2017), 150–51.
7. Wilson Bryan Key, *Subliminal Seduction: Ad Media's Manipulation of a Not So Innocent America* (Englewood Cliffs, N.J.: Prentice-Hall, 1973).
8. "Band Is Held Not Liable in Suicides of Two Fans," *The New York Times*, August 25, 1990, nytimes.com/1990/08/25/arts/band-is-held-not-liable-in-suicides-of-two-fans.html.
9. Charles Trappey, "A Meta-analysis of Consumer Choice and Subliminal Advertising," *Psychology and Marketing* 13, no. 5 (August 1996): 517–30, ir.nctu.edu.tw/bitstream/11536/14246/1/A1996UZ40600005.pdf.
10. Harold H. Kelley, "The Warm-Cold Variable in First Impressions of Persons," *Journal of Personality* 18, no. 4 (June 1950): 431–39, doi.org/10.1111/j.1467-6494.1950.tb01260.x.
11. K. S. Lashley, "The Problem of Serial Order in Behavior," in *Cerebral Mechanisms in Behavior: The Hixon Symposium*, ed. Lloyd A. Jeffress (New York: Wiley, 1951): 112–46.
12. Lowell H. Storms, "Apparent Backward Association: A Situational Effect," *Journal of Experimental Psychology* 55, no. 4 (April 1958): 390–95, doi:10.1037/h0044258.
13. S. J. Segal and C. N. Cofer, "The Effect of Recency and Recall on Word Association," *American Psychologist* 15, no. 7 (1960): 451.
14. The rundown of the key experiments in this paragraph comes from John A. Bargh and Tanya L. Chartrand, "The Mind in the Middle: A Practical Guide to Priming and Automaticity Research," in *Handbook of Research Methods in Social and Personality Psychology*, ed. Harry T. Reis and Charles M. Judd (New York: Cambridge University Press, 2000), 253–85, faculty.fuqua.duke.edu/~tlc10/bio/TLC_articles/2000/Bargh_Chartrand_2000.pdf.
15. A. N. Whitehead, *An Introduction to Mathematics* (London: Williams and Norgate, 1911), 45–6, Kindle edition (2012).
16. E. Tory Higgins, William S. Rholes, and Carl R. Jones, "Category Accessibility and Impression Formation," *Journal of Experimental Social Psychology* 13 (1977): 141–54, uni-muenster.de/imperia/md/content/psyifp/aeechterhoff/wintersemester2011-12/attitudesandsocialjudgment/higginsrholesjones_cataccessimpressform_jesp_1977.pdf.

17. Bargh and Chartrand, "Mind in the Middle," 256.
18. J. A. Bargh, M. Chen, and L. Burrows, "Automaticity of Social Behavior: Direct Effects of Trait Construct and Stereotype Activation on Action," *Journal of Personality and Social Psychology* 71, no. 2 (1996): 237, doi.org/10.1037/0022-3514.71.2.230.
19. Eugene M. Caruso et al., "Mere Exposure to Money Increases Endorsement of Free-Market Systems and Social Inequality," *Journal of Experimental Psychology: General* 142, no. 2 (2013), scholars.northwestern.edu/en/publications/mere-exposure-to-money-increases-endorsement-of-free-market-syste.
20. Ap Dijksterhuis and Ad van Knippenberg, "The Relation Between Perception and Behavior, or How to Win a Game of Trivial Pursuit," *Journal of Personality and Social Psychology* 74, no. 4 (1998): 865–77, psycnet.apa.org/record/1998-01060-003.
21. Johan C. Karremans, Wolfgang Stroebe, and Jasper Claus, "Beyond Vicary's Fantasies: The Impact of Subliminal Priming and Brand Choice," *Journal of Experimental Social Psychology* 42, no. 6 (November 2006): 792–98, sciencedirect.com/science/article/abs/pii/S0022103105001496.
22. Travis J. Carter, Melissa J. Ferguson, and Ran R. Hassin, "A Single Exposure to the American Flag Shifts Support Toward Republicanism up to 8 Months Later," *Psychological Science* 22, no. 8 (2011): 1011–18, doi.org/10.1177/0956797611414726.
23. Will M. Gervais and Ara Norenzayan, "Analytic Thinking Promotes Religious Disbelief," *Science* 336, no. 6080 (2012): 493–96, doi:10.1126/science.1215647.
24. Daniel Kahneman, *Thinking, Fast and Slow* (New York: Farrar, Straus and Giroux, 2011), 57.
25. Ran R. Hassin et al., "Subliminal Exposure to National Flags Affects Political Thought and Behavior," *Proceedings of the National Academy of Sciences of the United States of America* 104, no. 50 (December 2007): 19757–61, doi.org/10.1073/pnas.0704679104.
26. Jerry Kang, "Trojan Horses of Race," *Harvard Law Review* 118, no. 5 (March 2005): 1537, jstor.org/stable/4093447.
27. Kiju Jung et al., "Female Hurricanes Are Deadlier than Male Hurricanes," *Proceedings of the National Academy of Sciences of the United States of America*, June 2, 2014, doi:10.1073/pnas.1402786111.
28. Holly Yan, "Female Hurricanes Are Deadlier than Male Hurricanes, Study Says," CNN, September 1, 2016, cnn.com/2016/09/01/health/female-hurricanes-deadlier-than-male-hurricanes-trnd/index.html.
29. John Bargh, *Before You Know It: The Unconscious Reasons We Do What We Do* (New York: Atria Books, 2017), 108.
30. Bargh, *Before You Know It*, 105.
31. Lisa Zaval et al., "How Warm Days Increase Belief in Global Warming," *Nature Climate Change* 4 (February 2014): 143–47, doi.org/10.1038/nclimate2093.
32. Only two of the five studies had any connection to social priming. In one, the researchers had respondents complete a scrambled-word task

that sometimes included heat primes, sometimes included cold primes, and sometimes included neither. Participants exposed to heat primes were slightly (and statistically significantly) more likely to believe in global warming, and were more concerned about it, than those in the control or cold-priming groups. But the cold priming had no effect: participants in that group responded in a manner that was statistically indistinguishable from members of the control group. In another study in the article, the researchers found that using "global warming" (which "may prime heat-related cognitions") versus "climate change" made no difference in respondents' views—a result that has been replicated elsewhere more recently.

33. See, for example, Kahan's remarks in Umair Irfan, "People Furthest Apart on Climate Views Are Often the Most Educated," *Scientific American*, August 22, 2017, scientificamerican.com/article/people-furthest-apart-on-climate-views-are-often-the-most-educated.
34. Bargh, *Before You Know It*, 199.
35. Matt DeLisi, "Broken Windows Works," *City Journal*, May 29, 2019, city-journal.org/broken-windows-policing-works.
36. Bernard E. Harcourt and Jens Ludwig, "Broken Windows: New Evidence from New York City and a Five-City Social Experiment," *Public Law and Legal Theory*, no. 93 (2005), chicagounbound.uchicago.edu/public_law_and_legal_theory/48.
37. See David Kennedy's work, for example, which suggests that small groups of young men are responsible for much of the gunplay in violent neighborhoods. David M. Kennedy, *Don't Shoot: One Man, a Street Fellowship, and the End of Violence in Inner-City America* (New York: Bloomsbury USA, 2011).
38. Matthew Desmond, Andrew V. Papachristos, and David S. Kirk, "Police Violence and Citizen Crime Reporting in the Black Community," *American Sociological Review* 81, no. 5 (October 2016): 857–76, doi.org/10.1177/0003122416663494.
39. Bargh, *Before You Know It*, 208.
40. Tom Stafford, "The Perspectival Shift: How Experiments on Unconscious Processing Don't Justify the Claims Made for Them," *Frontiers in Psychology*, September 19, 2014, frontiersin.org/articles/10.3389/fpsyg.2014.01067/full.
41. Daryl J. Bem, "Feeling the Future: Experimental Evidence for Anomalous Retroactive Influences on Cognition and Affect," *Journal of Personality and Social Psychology* 100, no. 3 (2011): 407–25, ncbi.nlm.nih.gov/pubmed/21280961.
42. Paul Jump, "A Star's Collapse," *Inside Higher Ed*, November 28, 2011, insidehighered.com/news/2011/11/28/scholars-analyze-case-massive-research-fraud.
43. Mieke Verfaellie and Jenna McGwin, "The Case of Diederik Stapel," *Psychological Science Agenda*, December 2011, apa.org/science/about/psa/2011/12/diederik-stapel.aspx.
44. Shannon Palus, "Diederik Stapel Now Has 58 Retractions," Retraction Watch, last modified December 8, 2015, retractionwatch.com/2015/12/08/diederik-stapel-now-has-58-retractions.

45. Tal Yarkoni, "The Psychology of Parapsychology, or Why Good Researchers Publishing Good Articles in Good Journals Can Still Get It Totally Wrong," TalYarkoni.org, talyarkoni.org/blog/2011/01/10/the-psychology-of-parapsychology-or-why-good-researchers-publishing-good-articles-in-good-journals-can-still-get-it-totally-wrong.
46. Chris Chambers, *The Seven Deadly Sins of Psychology: A Manifesto for Reforming the Culture of Scientific Practice* (Princeton, N.J.: Princeton University Press, 2017), 23.
47. Leslie K. John, George Loewenstein, and Drazen Prelec, "Measuring the Prevalence of Questionable Research Practices with Incentives for Truth Telling," *Psychological Science* 23, no. 5 (2012): 524–32, doi:10.1177/0956797611430953.
48. Chambers, *The Seven Deadly Sins of Psychology*, 3.
49. Joseph P. Simmons, Leif D. Nelson, and Uri Simonsohn, "False-Positive Psychology: Undisclosed Flexibility in Data Collection and Analysis Allows Presenting Anything as Significant," *Psychological Science* 22, no. 11 (November 2011): 1359–66, doi.org/10.1177/0956797611417632.
50. Tom Chivers, "What's Next for Psychology's Embattled Field of Social Priming," *Nature*, December 11, 2019, nature.com/articles/d41586-019-03755-2.
51. Kahneman, *Thinking, Fast and Slow*, 57.
52. Dan Goldstein, "Kahneman on the Storm of Doubts Surrounding Social Priming Research," *Decision Science News*, October 5, 2012, decisionsciencenews.com/2012/10/05/kahneman-on-the-storm-of-doubts-surrounding-social-priming-research.
53. Stéphane Doyen et al., "Behavioral Priming: It's All in the Mind, but Whose Mind?," *Plos One* 7, no. 1 (January 2012), doi.org/10.1371/journal.pone.0029081.
54. Colin F. Camerer et al., "Evaluating the Replicability of Social Science Experiments in *Nature* and *Science* between 2010 and 2015," *Nature Human Behaviour* 2, no. 9 (2018): 637–44, doi.org/10.1038/s41562-018-0399-z; Brian Resnick, "More Social Science Studies Just Failed to Replicate. Here's Why This is Good," *Vox*, August 27, 2018, vox.com/science-and-health/2018/8/27/17761466/psychology-replication-crisis-nature-social-science.
55. Michael O'Donnell et al., "Registered Replication Report: Dijksterhuis and van Knippenberg (1998)," *Perspectives on Psychological Science* 13, no. 2 (March 2018): 268–94, doi:10.1177/1745691618755704.
56. Richard A. Klein et al., "Investigating Variation in Replicability," *Social Psychology* 45, no. 3 (January 2014): pp. 142–152, https://doi.org/10.1027/1864-9335/a000178, 149.
57. Gary Smith, "Hurricane Names: A Bunch of Hot Air?," *Weather and Climate Extremes* 12 (June 2016): 80–84, doi.org/10.1016/j.wace.2015.11.006.
58. Brian Nosek et al., "Estimating the Reproducibility of Psychological Science," *Science* 349, no. 6251 (August 2015), science.sciencemag.org/content/349/6251/aac4716.
59. Richard A. Klein et al., "Many Labs 2: Investigating Variation in Replicability Across Sample and Setting," *Advances in Methods and Practices in Psychological Science* 1, no. 4 (2018): 443–90, osf.io/8cd4r/.

60. Chivers, "What's Next for Psychology's Embattled Field of Social Priming."
61. Alexander A. Aarts et al., "Estimating the Reproducibility of Psychological Science," *Science* 349, no. 6251 (August 2015), science.sciencemag.org/content/349/6251/aac4716/tab-pdf.
62. Ulrich Schimmack and Jerry Brunner, "Z-Curve: A Method for the Estimating Replicability Based on Test Statistics in Original Studies," OSF Preprints, November 17, 2017, doi:10.31219/osf.io/wr93f.
63. Ulrich Schimmack, "'Before You Know It' by John A. Bargh: A Quantitative Book Review," *Replicability-Index* (blog), November 28, 2017, replicationindex.com/2017/11/28/before-you-know-it-by-john-a-bargh-a-quantitative-book-review.
64. Daniel T. Gilbert et al., "Comment on 'Estimating the Reproducibility of Psychological Science,'" *Science* 351, no. 6277 (March 2016): 1037, doi.org/10.1126/science.aad7243.
65. Jonathon Keats, "Debating Psychology's Replication Crisis," *Discover*, August 25, 2016, discovermagazine.com/mind/debating-psychologys-replication-crisis.
66. "28 Classic and Contemporary Psychology Findings Replicated in More than 60 Laboratories Each Across Three Dozen Nations and Territories," *Center for Open Science*, November 19, 2018, cos.io/about/news/28-classic-and-contemporary-psychology-findings-replicated-more-60-laboratories-each-across-three-dozen-nations-and-territories.
67. "Preregistration," *Center for Open Science*, n.d., cos.io/our-services/prereg.
68. I ran down some of the issues with IRBs here: Jesse Singal, "Is a Portland Professor Being Railroaded by His University for Criticizing Social-Justice Research?," *New York*, January 11, 2019, nymag.com/intelligencer/2019/01/is-peter-boghossian-getting-railroaded-for-his-hoax.html.
69. "Registered Reports: Peer Review Before Results Are Known to Align Scientific Values and Practices," Center for Open Science, cos.io/rr/.
70. Aubrey Clayton, "The Flawed Reasoning Behind the Replication Crisis," *Nautilus* 74 (August 2019), nautil.us/issue/74/networks/the-flawed-reasoning-behind-the-replication-crisis.
71. $P(A/B) = (P[B/A] \times P[A]) \ / \ P(B)$, where $A$, $B$ = events; $P(A|B)$ = probability of $A$ given $B$ is true; $P(B|A)$ = probability of $B$ given $A$ is true; $P(A)$, $P(B)$ = independent probabilities of $A$ and $B$.
72. Clayton, "Flawed Reasoning Behind the Replication Crisis."
73. For the record, the full calculation, Clayton writes, is $(12\,p) \times (1/1{,}000)$ v. $(p) \times (999/1{,}000)$.
74. Eric-Jan Wagenmakers et al., "Why Psychologists Must Change the Way They Analyze Their Data: The Case of Psi: Comment on Bem (2011)," *Journal of Personality and Social Psychology* 100, no. 3 (2011): 426–32, doi:10.1037/a0022790.
75. Lee Billings, *Scientific American*, June 15, 2017, scientificamerican.com/article/china-shatters-ldquo-spooky-action-at-a-distance-rdquo-record-preps-for-quantum-internet.
76. Chivers, "What's Next for Psychology's Embattled Field of Social Priming."

## 8. NUDGING AHEAD

1. "District Attorney Vance, Commissioner Bratton, Mayor de Blasio Announce New Structural Changes to Criminal Summonses Issued in Manhattan," Manhattan District Attorney's Office, March 1, 2016, manhattanda.org/district-attorney-vance-commissioner-bratton-mayor-de-blasio-announce-new-structural-c.
2. Brice Cooke et al., "Using Behavioral Science to Improve Criminal Justice Outcomes: Preventing Failures to Appear in Court," ideas42 and UChicago Urban Labs (January 2018): 4, ideas42.org/wp-content/uploads/2018/03/Using-Behavioral-Science-to-Improve-Criminal-Justice-Outcomes.pdf.
3. "NYC Summons Redesign," ideas42.org, ideas42.org/wp-content/uploads/2018/01/I42–954_SummonsForm_exp_3-1.pdf.
4. Cooke et al., "Using Behavioral Science to Improve Criminal Justice Outcomes: Preventing Failures to Appear in Court."
5. Gina Martinez, "Summons Recipients to Receive Text Reminders," *QNS*, February 3, 2018, qns.com/story/2018/02/03/summons-recipients-to-receive-text-reminders/.
6. Richard H. Thaler and Cass R. Sunstein, *Nudge: Improving Decisions About Health, Wealth, and Happiness*, 2nd ed. (New York: Penguin Books, 2009), 6.
7. Katrin Bennhold, "Britain's Ministry of Nudges," *The New York Times*, December 7, 2013, nytimes.com/2013/12/08/business/international/britains-ministry-of-nudges.html.
8. Thaler, interview by Stephen J. Dubner, July 11, 2018, freakonomics.com/podcast/richard-thaler/; Alain Samson, "An Introduction to Behavioral Economics," BehavioralEconomics.com, behavioraleconomics.com/resources/introduction-behavioral-economics/.
9. Royal Swedish Academy of Sciences, "The Prize in Economic Sciences 2017," press release, October 9, 2017, nobelprize.org/prizes/economic-sciences/2017/press-release.
10. Thaler and Sunstein, *Nudge*, 6.
11. Michael Lewis, *The Undoing Project: A Friendship That Changed Our Minds* (New York: W. W. Norton, 2016), 147.
12. Lewis, *The Undoing Project*, 148.
13. Lewis, *The Undoing Project*, 157, 177, 177–78.
14. "Amos Tversky, Leading Decision Researcher, Dies at 59," Stanford News Service, June 5, 1996, news.stanford.edu/pr/96/960605tversky.html.
15. "The Sveriges Riksbank Prize in Economic Sciences in Memory of Alfred Nobel 2002," press release, October 9, 2002, nobelprize.org/prizes/economic-sciences/2002/press-release/.
16. Amos Tversky and Daniel Kahneman, "Extensional versus Intuitive Reasoning: The Conjunction Fallacy in Probability Judgment," *Psychological Review* 90, no. 4 (October 1983): 297, doi.org/10.1037/0033-295X.90.4.293.
17. Amos Tversky and Daniel Kahneman, "The Framing of Decisions and the Psychology of Choice," *Science* 211, no. 4481 (January 1981): 453, doi.org/10.1126/science.7455683.

18. Cooke et al., "Using Behavioral Science to Improve Criminal Justice Outcomes: Preventing Failures to Appear in Court," 5.
19. Cass R. Sunstein, *Simpler: The Future of Government* (New York: Simon & Schuster, 2013), 51.
20. Sunstein, *Simpler*, 52.
21. "Richard H. Thaler: Facts," Nobel Prize, nobelprize.org/prizes/economic-sciences/2017/thaler/facts.
22. Michael Lewis, "Obama's Way," *Vanity Fair*, October 2012, vanityfair.com/news/2012/10/michael-lewis-profile-barack-obama.
23. James Surowiecki, "A Smarter Stimulus," *The New Yorker*, January 26, 2009, newyorker.com/magazine/2009/01/26/a-smarter-stimulus.
24. Jonathan Weisman and Jess Bravin, "Obama's Regulatory Czar Likely to Set a New Tone," *The Wall Street Journal*, January 8, 2009, wsj.com/articles/SB123138051682263203.
25. Office of the Press Secretary, the White House, "Executive Order—Using Behavioral Science Insights to Better Serve the American People," press release, September 15, 2015, obamawhitehouse.archives.gov/the-press-office/2015/09/15/executive-order-using-behavioral-science-insights-better-serve-american.
26. Press Secretary, the White House, "Executive Order—Using Behavioral Science Insights to Better Serve the American People."
27. U.S. Executive Office of the President–National Science and Technology Council, *Social and Behavioral Sciences Team: Annual Report* (Washington, D.C.: U.S. Executive Office of the President–National Science and Technology Council, 2015), obamawhitehouse.archives.gov/sites/default/files/microsites/ostp/sbst_2015_annual_report_final_9_14_15.pdf.
28. Danny Vinik, "Obama's Effort to 'Nudge' America," *Politico*, October 15, 2015, politico.com/agenda/story/2015/10/obamas-effort-to-nudge-america-000276/.
29. Rosa Li, "The Other Essential Pandemic Office Trump Eliminated," *Slate*, March 18, 2020, slate.com/technology/2020/03/coronavirus-social-behavior-trump-white-house.html.
30. Economist Staff, "Nudge Comes to Shove," *Economist*, May 18, 2017, economist.com/international/2017/05/18/policymakers-around-the-world-are-embracing-behavioural-science.
31. Julia Kagan, "Pension Protection Act of 2006," *Investopedia*, last modified April 3, 2019, investopedia.com/terms/p/pensionprotectionact2006.asp.
32. Jeffrey W. Clark, Stephen P. Utkus, and Jean A. Young, *Automatic Enrollment: The Power of the Default*, Vanguard Research, February 2018, 4.
33. Robert M. Bond et al., "A 61-Million-Person Experiment in Social Influence and Political Mobilization," *Nature* 489, no. 7415 (September 2012): 295–98, doi:10.1038/nature11421.
34. Romain Cadario and Pierre Chandon, "Which Healthy Eating Nudges Work Best? A Meta-analysis of Field Experiments," *Marketing Science* 39, no. 3 (2020): 465–86, doi.org/10.1287/mksc.2018.1128.

35. Richard A. Klein et al., "Investigating Variation in Replicability," *Social Psychology* 45, no. 3 (January 2014): pp. 142–152, https://doi.org/10.1027/1864-9335/a000178, 147.
36. Adam Arshad, Benjamin Anderson, and Adnan Sharif, "Comparison of Organ Donation and Transplantation Rates between Opt-Out and Opt-In Systems," *Clinical Investigation* 95, no. 6 (June 2019): 1453, doi.org/10.1016/j.kint.2019.01.036.
37. Brian H. Willis and Muireann Quigley, "Opt-Out Organ Donation: On Evidence and Public Policy," *Journal of the Royal Society of Medicine* 107, no. 2 (February 2014): 56–60, doi:10.1177/0141076813507707.
38. Wokje Abrahamse and Rachael Shwom, "Domestic Energy Consumption and Climate Change Mitigation," *WIREs Climate Change* 9, no. 4 (July/August 2018), doi.org/10.1002/wcc.525.
39. Philip Oreopoulos and Uros Petronijevic, "The Remarkable Unresponsiveness of College Students to Nudging and What We Can Learn from It," National Bureau of Economic Research Working Papers, no. 26059 (July 2019), nber.org/papers/w26059.
40. Kelli A. Bird et al., "Nudging at Scale: Experimental Evidence from FAFSA Completion Campaigns," National Bureau of Economic Research Working Papers, no. 26158 (August 2019), nber.org/papers/w26158.
41. *Check in the Mail or More in the Paycheck: Does the Effectiveness of Fiscal Stimulus Depend on How It Is Delivered?*, prepared by Claudia R. Sahm, Matthew D. Shapiro, and Joel Slemrod in cooperation with the Federal Reserve Board (Washington, D.C., 2010), federalreserve.gov/pubs/feds/2010/201040/201040pap.pdf.
42. Leigh Giangreco, "How Behavioral Science Solved Chicago's Plastic Bag Problem," *Politico*, November 2019, politico.com/news/magazine/2019/11/21/plastic-bag-environment-policy-067879.
43. Robert Kuttner, "The Radical Minimalist: Obama's New Regulatory Czar Likes Market Incentives. Will They Work?," *The American Prospect*, April 2009: pp. 28–30, prospect.org/features/radical-minimalist.
44. Tom Slee, *No One Makes You Shop at Wal-Mart: The Surprising Deceptions of Individual Choice* (Toronto: Between the Lines, 2006).
45. Samuel Moyn, "The Nudgeocrat," *The Nation*, June 17–24, 2019, thenation.com/article/archive/cass-sunstein-on-freedom-book-review.
46. William H. Simon, "The Republic of Choosing," *Boston Review*, July 8, 2013, bostonreview.net/us-books-ideas/cass-sunstein-simpler-future-government-republic-choosing.
47. "CFPB Finalizes 'Know Before You Owe' Mortgage Forms," Consumer Financial Protection Bureau, November 20, 2013, consumerfinance.gov/about-us/newsroom/cfpb-finalizes-know-before-you-owe-mortgage-forms/.
48. Simon, "The Republic of Choosing."
49. Robert H. Frank, *Under the Influence: Putting Peer Pressure to Work* (Princeton, N.J.: Princeton University Press, 2020), 215.
50. David V. Johnson, "Twilight of the Nudges," *New Republic*, October 27, 2016, newrepublic.com/article/138175/twilight-nudges.

51. "The Criminal Justice Reform Act: One Year Later," New York City Council, 2017, council.nyc.gov/the-criminal-justice-reform-act-one-year-later.
52. For the truly, obsessively inquisitive reader, this would be NYPD summons number 0196 441 594, issued August 17, 2019.

## CONCLUSION: ESCAPE FROM PRIMEWORLD

1. Charles Tilly, *Why? What Happens When People Give Reasons . . . and Why* (Princeton, N.J.: Princeton University Press, 2006), 64, 65.
2. Nina Strohminger, "Author Reply: Grasping the Nebula: Inelegant Theories for Messy Phenomena," *Emotion Review* 6, no. 3 (July 2014): 226, doi.org/10.1177/1754073914524455.
3. Hart Blanton and Elif G. Ikizer, "The Bullet-Point Bias: How Diluted Science Communications Can Impede Social Progress," in *The Politics of Social Psychology*, ed. Jarret T. Crawford and Lee Jussim (New York: Routledge, 2018), 169.
4. Blanton, "Bullet-Point Bias," 169.
5. Anand Giridharadas, *Winners Take All: The Elite Charade of Changing the World* (New York: Knopf, 2018), 30.
6. Gregory M. Walton, "The New Science of Wise Psychological Interventions," *Current Directions in Psychological Science* 23, no. 1 (2014): 73–82, doi:10.1177/0963721413512856.
7. Blanton, "Bullet-Point Bias," 180.
8. Blanton, "Bullet-Point Bias," 180–81.
9. Natalie M. Daumeyer et al., "Consequences of Attributing Discrimination to Implicit vs. Explicit Bias," *Journal of Experimental Social Psychology* 84 (September 2019): 1–10, doi.org/10.1016/j.jesp.2019.04.010.
10. Hans IJzerman et al., "Is Social and Behavioural Science Evidence Ready for Application and Dissemination?," preprint, 2020, doi.org/10.31234/osf.io/whds4.

# ACKNOWLEDGMENTS

There are too many people to thank! Thanks again, Mom and Dad. For everything. Also on the family front, I'm lucky to have two great (and, to use slightly outdated language, ball-busting) brothers, Gabe and Alex, both of whom, it is important to note, are slightly shorter than I am; two living grandmothers I hold very dear (and don't call enough), Shavee Altman and Edith Singal-Katz; and an endlessly fascinating and loving uncle, Robert Altman. I love all of you. I hope this book honors the memories of Shavee's husband, Harry Altman, who continues to be mourned by three generations, and Bruce's father, Richard Singal—a small part of me will always be missing for having never met him, but he continues to live on through his son, and through the photos in which he is boxing and drumming and looking both cooler and more dapper than I ever will.

Andrew Stuart came up with the initial idea for *The Quick Fix*. It really wouldn't exist without him, and his help shaping the proposal made a huge difference. I simply couldn't have asked for a better agent. But I could say the same thing about my editor at FSG: Alex Star has brought what was at first a fairly rough manuscript along masterfully, expertly guiding me through difficult questions about what to emphasize and what to cut, how to boil down complicated debates for a lay audience, and kindly forgiving my tendency to recycle certain phrases (like "zooming out") a bit too frequently. Isaac Scher did a sensational job fact-checking under tight time pressure and frazzled supervision, and I'm not sure how I would have finished the book without him.

Zooming out, I've had some great editors over the years, and their fingerprints are all over this book: in particular, Peter Canellos at *The Boston Globe*, Jesse Wegman at *The Daily Beast*, and Jebediah Reed at *New York* magazine (an outlet that gave me a true dream job that led directly to this book). I learned a great deal from them, and the consistent messages I received from all three—delivered with varying degrees of exasperation, as the situation warranted—were to shy from certitude, to question my first impressions, and to always investigate further. You may have noticed those themes popping up once or twice in the preceding pages.

I owe a debt to every expert who appears in this book to help elucidate a concept, cast doubt upon a questionable claim, or translate baffling complexity into plainspeak, whether or not I interacted with them directly. But a number of individuals have been particularly generous with their time over the years and deserve recognition for answering many of my questions, reading drafts of various passages or chapters, or both. It would be impossible to list all of them (and surely I am forgetting some, for which I apologize), but, among others, I'd like to thank Hart Blanton, Nick Brown, Tom Chivers, Aubrey Clayton, Marcus Crede, Daniel DeFraia, Patrick Forscher, Andrew Gelman, Jane Gillham, Roger Giner-Sorolla, Patricia Resick, Stuart Ritchie, Ulrich Schimmack, Joseph Simmons, Uri Simonsohn, Will Storr, and Carol Tavris.

I've been improbably fortunate to have made such kind, talented, funny, and down-to-earth friends during my various stops over the years. I'd like to thank and recognize and in some cases lightheartedly ridicule my closest hometown friends, Andrew Bielak, Dan Bromfield, Jesse Feinberg, and Denis Vidal (an honorary but equal-status member), as well as fellow beloved Newtonites Molly Cosgrove, Sarah Rastegar, and Danielle Wessler; Randolph Brickey from Brandeis; Elsie Bell and Joanna Geller—both of whom serve as north stars for me on anything having to do with kids, inequality, or both—from the University of Michigan; Thomas Coen and David Spett from my D.C. years; Emily Staudenmaier and Tess Varney from William Street in Cambridge; Ezra Levin, Rob Rosenbaum, Greg Rosalsky, and so many others from my truly excellent class at the Policy School Formerly Known as Woodrow Wilson (I will see all of you soon at D-Bar); Riley Ohlson and Christina Tsafoulias from Berlin (danke schön, Robert Bosch Foundation); Ben Adler, Nick Clairmont, Dave Cole, Carmen Hilbert (slight leap of faith), Matt Kassel, Alice Robb, and Jordan

Smith from Brooklyn; and my trusty podcast cohost, Katie Herzog. I'm extremely fortunate to have all of you in my life.

Some sad parting words: In 2018, Dan Bromfield's wife, Eliza Forrest Kaye Bromfield, passed away suddenly and unexpectedly at the age of thirty. To adequately describe what a wonderful and kind and glowing person she was would require a whole book of its own, and it's unfathomably unfair that Dan—and the rest of us—lost her. It's on all of us to live up to the blazing example she set.

As I was making my very final tweaks to this book, my mom, Sydney Altman, was diagnosed with stage IV lung cancer. While her initial response to treatment has been favorable, at least given the circumstances, she and the rest of my family face a very difficult, terrifying road. Mom: I have been blessed with far more than I deserve, and I owe so much of it to you—to your insatiable curiosity, your sense of humor, and your arch skepticism of puffed-up authority. I worry that sometimes you lose sight of this, but I plan on reminding you over and over, for as long as I can. I love you.

# INDEX

Page numbers in *italics* refer to illustrations.

A NOTE ABOUT THE AUTHOR

Jesse Singal is a contributing writer at *New York* and the former editor of the magazine's *Science of Us* online vertical, as well as the cohost of the podcast *Blocked and Reported*. His work has appeared in *The New York Times*, *The Atlantic*, *Slate*, *The Boston Globe*, and other outlets. He is a former Robert Bosch Foundation fellow in Berlin and holds a master's degree from Princeton University's School of Public and International Affairs.